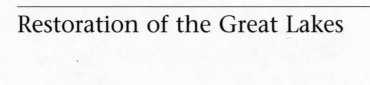

Restoration of the Great Lakes

Mark Sproule-Jones

Restoration of the Great Lakes:
Promises, Practices, Performances

UBCPress · Vancouver · Toronto

09 08 07 06 05 04 03 02   5 4 3 2 1

Printed in Canada on acid-free paper

National Library of Canada Cataloguing in Publication Data

Sproule-Jones, Mark, 1941-
  Restoration of the Great Lakes

  Includes bibliographical references and index.
  ISBN 0-7748-0870-5

  1. Environmental protection – Great Lakes. 2. Water – Pollution – Great Lakes. 3. Conservation of natural resources – Government policy – Great Lakes. 4. Great Lakes – Environmental conditions. 5. Lake renewal – Great Lakes. I. Title.
TD223.3.S67 2002   333.91'63153'0977   C2002-910148-4

Canadä

UBC Press gratefully acknowledges the financial support for our publishing program of the Government of Canada through the Book Publishing Industry Development Program (BPIDP), and of the Canada Council for the Arts, and the British Columbia Arts Council.

This book has been published with the help of a grant from the Humanities and Social Sciences Federation of Canada, using funds provided by the Social Sciences and Humanities Research Council of Canada.

UBC Press
The University of British Columbia
2029 West Mall
Vancouver, BC V6T 1Z2
604-822-5959 / Fax: 604-822-6083
www.ubcpress.ca

To Vincent and Lin:
confidants, friends, mentors, teachers

Contents

Figures and Tables

Acknowledgments

Many people should be thanked for contributing to the research, writing, and publication of this book, not least of which are my friends and colleagues who put up with its lengthy formative years. I would like to acknowledge the help of the public servants who provided the information, reports, and comments upon which much of this book relies. They perhaps will be disappointed in its conclusions but will be resilient enough to recognize how self-governance and the restoration of the Great Lakes cannot simply be a product of bureaucratic hard work. There were other stakeholders around the lakes who also contributed much, and the Hamilton Harbour Remedial Action Plan contributed more than a little to my professional and personal life.

Many academic colleagues helped polish the research and writing throughout five years of work. I would like to acknowledge April Beaulé, Robert Bish, Barbara Carroll, David Feeny, Tim Heinmiller, Tim Hennessey, Carolyn Johns, Trish Johnson, Réjean Landry, Catherine McLinden, Robert Paehlke, Margaret Polski, Lin Ostrom, Vincent Ostrom, Filippo Sabetti, Jon Sutinen, Karen Thomas, and Michael Trebilcock. Finally, I wish to thank Lori Ewing and Deanna Goral for manuscript preparation and the Tri-Council's EcoResearch Program for Hamilton Harbour for funding this and companion research studies.

Acronyms

AOCs	Areas of Concern
COA	Canada-Ontario Agreement
CPR	common-pool resources; common-property resources
EPA	Environmental Protection Agency
GLWQA	Great Lakes Water Quality Agreement
GLWQB	Great Lakes Water Quality Board
GLWQI	Great Lakes Water Quality Initiative
IAD	institutional analysis and development
IJC	International Joint Commission
LaMP	Lakewide Management Plan
PAH	polycyclic aromatic hydrocarbon
PCB	polychlorinated biphenyl
RAB	research advisory board
RAP	remedial action plan

Restoration of the Great Lakes

1
Introduction

The world has seen many major experiments. The changes brought about by the American, French, and Russian revolutions are good examples of social experiments on a grand scale. But small experiments go on daily – experiments in lifestyles, childrearing, and architectural design. This book is about experiments that lie somewhere between large and small events. They are occurring on the North American Great Lakes and consist of designing institutions to restore degraded environmental sites throughout the lakes.

Few people would consider the Great Lakes to be small or unimportant. The five lakes occupy an area greater than half a billion square kilometres, have a shoreline of 17,000 kilometres, and are surrounded by a population of over 33 million. The waters themselves exceed 22,000 cubic kilometres. For centuries they have provided sustenance and carrying capacity for a range of human, animal, and plant life. Lakes Superior, Michigan, Huron, Erie, and Ontario constitute about one-fifth of the world's fresh surface-water supply (US Environmental Protection Agency and Government of Canada 1995). So what happens to the Great Lakes is indeed of major significance.

Since 1985, many small experiments in institutional design have occurred all over the Great Lakes. The national governments of Canada and the United States, the governments of the states and provinces bordering the lakes, and the International Joint Commission (established in 1909 by Canada and the United States to study and recommend solutions to transborder problems) have designed and are designing institutions to remedy the severely degraded environments of 43 bays, harbours, and river mouths along the lakes' shorelines. These new institutional designs for the 43 Areas of Concern (AOCs) recognize, in part, that the traditional institutional designs for environmental cleanup are inadequate. Users and regulators of the lakes simply neglected to sustain a range of human and non-human uses of the resource. Living systems were dying, often slowly, while public authorities failed in their mandates. Some new experiments were essential.

Promises

The restoration experiments for the 43 AOCs promise much. First, they promise a way in which resource users, regulators, and those with an interest in a clean local environment can work together to restore beneficial uses in the affected areas. They present a method for establishing efficient resource use, whereby the values of the alternative uses of each site can be balanced and weighed by the stakeholders involved.

The experiments in institutional design will ensure more than efficiency. They promise to empower local stakeholders to fashion their own solutions to resource degradation. All that must be kept in mind is the goal: restoring the beneficial uses that environmental scientists have designated (after many years of study) as having been severely impaired. Local stakeholders could indeed become architects of their own futures.

Third, the experiments can facilitate avenues for collaboration and cooperation among the various groups and stakeholders, who are sometimes at odds. Governments must work with private interests, environmentalists with polluters, fishers with shipping interests, industries with municipalities, and so on. It is possible that a sense of democratic civility and cooperation can be engendered through establishing new ties and ventures among disparate interests.

Finally, the experiments provide a way to measure and assess the accountability of public agencies. Restoration of beneficial uses in the Great Lakes is the standard of performance. The means to get there includes the coordinated efforts of public agencies as regulators and implementers of this public policy. The efforts can be evaluated and the accountabilities of agencies assessed. The national, state, provincial, and local governments, as well as Native communities, stand to gain from this exercise of democracy in action.

Targets

In 1985, the International Joint Commission (IJC) asked the governments of Canada and the United States to formulate remedial action plans (RAPs) for each of the 43 AOCs. No framework for institutional design was presented, other than a request that local stakeholders should be involved and that resource interdependencies should be recognized within an ecosystem planning approach. RAPs were to identify problems and solutions in the sites (Stage 1), develop implementation strategies (Stage 2), and report on the successful restoration of beneficial uses (Stage 3). RAPs represent a third wave of efforts by the governments of the United States and Canada to clean up the Great Lakes.

In 1972, Canada and the United States signed the first Great Lakes Water Quality Agreement (GLWQA) to restore and maintain the water quality of the Great Lakes. At the time, the focus was on rescuing Lake Erie from severe eutrophication and, ostensibly, its "death" (Ashworth 1986, 123-24). Major expenditures on sewage treatment plants and some restructuring of manufacturing processes successfully reduced the phosphorous inputs. Phosphorous had acted to stimulate algae and plant growth in Lake Erie, the shallowest of the Great Lakes.

The 1970s were a period during which an awareness of the range and severity of toxic contaminants entered the public agenda, and the GLWQA was amended in 1978 to include toxic substance loadings. It soon became apparent that these new concerns presented more serious and intractable ecosystem problems than did the phosphorus. The sources, pathways, and consequences of many heavy metals and persistent organic compounds still remain uncertain. Alleviating this problem would entail a more complex program than the nutrient reduction program of the 1970s.

In 1985, the federal and state/provincial governments recognized that the development of something like RAPs was needed to restore impaired beneficial uses in the 43 AOCs. The Water Quality Board, a principal advisory group to the International Joint Commission, listed 14 impaired beneficial uses. This list was later incorporated into the protocol of a revised GLWQA in 1987. The list is curious in many ways. It did not, and does not, include all impaired beneficial uses for the Great Lakes, particularly human uses such as water contact recreation or restored commercial fisheries. Nor did the list specify what the key concepts – such as "degradation of fish and wildlife populations" – meant operationally. The list of use impairments is shown in Figure 1.1; the 43 AOCs are shown in Figure 1.2; and the specific use impairments for each AOC are catalogued in Table 1.1.

Of the original 43 AOCs, 24 were in US waters, 12 in Canadian waters, and 7 in jointly shared "passageways."

The GLWQA also requires that states and provinces embody in their RAPs "a systematic and comprehensive ecosystem approach to restoring and protecting beneficial uses in Areas of Concern" (Great Lakes Water Quality Agreement of 1978, International Joint Commission 1994, Annex 2, 2[a]). No prescriptions were made with regard to

Figure 1.1

Use impairments for Areas of Concern on the Great Lakes

- Restrictions on fish and wildlife consumption
- Tainting of fish and wildlife flavour
- Degradation of fish and wildlife populations
- Fish tumours or other deformities
- Bird or animal deformities or reproduction problems
- Degradation of benthos
- Restrictions on dredging activities
- Eutrophication or undesirable algae
- Restriction on drinking water consumption, or taste and odour problems
- Beach closings
- Degradation of aesthetics
- Added costs to agriculture or industry
- Degradation of phytoplankton and zooplankton populations
- Loss of fish and wildlife habitat

operationalizing the ecosystem approach. Some early commentators eagerly interpreted the ecosystem approach as one that involved and considered "all users in policy making and management, including scientists, regulators, industry, citizen representatives, and others" (Hartig and Zarull 1992, 13). The IJC was more sanguine in recognizing that each AOC had diverse interests and that RAPs were simply "a step toward ecosystem management" (IJC 1991a, 3).

Multiple Uses and Multiple Problems
The Great Lakes, like all other major waterways, support a variety of human and non-human uses. Many uses tend to be centred on harbours, bays, and rivers, in part because human settlements cluster around these more productive sites. What uses do we normally find? The lakes were originally opened up as navigation and transportation channels, and shipping remains a major, though declining, use. It should be noted that shipping does not have relate exclusively to ships: many rivers in the Great Lakes were major carriers of logs, with booms floating downstream from the logging camps. A number of recreational and commercial food fisheries still exist, and habitats for a range of wildlife and fish remain scattered throughout the lakes. The waters themselves are used for domestic, industrial, and (occasionally) agricultural "consumption." Recreational uses include not just pleasure boating and fishing and hunting, but also contact sports and swimming. Last, but certainly not least, the lakes have proved to be cheap places for dumping wastes from homes, communities, and industries. Structures like the dedicated coal pier in

Figure 1.2

Areas of Concern in the Great Lakes–St. Lawrence River basin

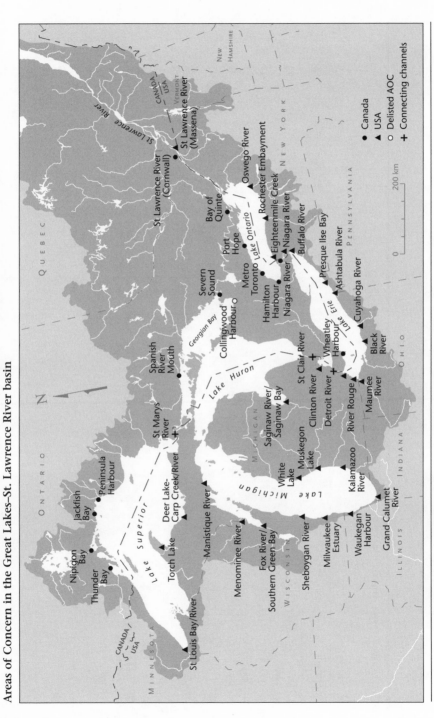

Source: Environment Canada, *Our Great Lakes*, 1999, <www.ec.on.gc.ca/glimr/maps-e.html>.

Table 1.1

Categories of use impairments for Areas of Concern on the Great Lakes

Area of Concern	Ecological health and reproduction	Habitat	Human health	Human use/welfare
Lake Superior				
Peninsula Harbour	3, 6	14	1	7, 9
Jackfish Bay	3, 4, 5?, 6	14	1?	7, 11
Nipigon Bay	3, 4?, 6, 8	14	1	7, 11
Thunder Bay	3, 4, 5?, 6, 13	14	1, 10, 2	7, 11, 12
St. Louis Bay/River	3, 4, 5?, 6	14	1, 10, 2?	7, 11
Torch Lake				
Deer Lake-Carp Creek/River	6		1	
Lake Michigan				
Manistique River	3, 6	14	1	7, 11
Menominee River	3, 6	14	1	7, 11
Fox River/Southern Green Bay	3, 4?, 5, 6, 8, 13	14	1, 10, 2?	7, 9, 11
Sheboygan River	3, 4, 5, 6, 8, 13	14	1, 10	7, 11
Milwaukee Estuary	3?, 4, 5, 6, 8, 13	14	1, 10	7, 11
Waukegan Harbor	3?, 5?, 6, 13	14	1, 10, 2?	7, 9?, 11
Grand Calumet River/Indiana Harbor Canal	3, 4, 5, 6, 8, 13	14	1, 10, 2	7, 9, 11, 12
Kalamazoo River	3, 5?	14	1	11
Muskegon Lake	3, 5?, 6, 8, 13?	14	1	7, 9, 11
White Lake	3, 5?, 6, 8, 13?	14	1	7, 9, 11

Lake Huron															
Saginaw River/Bay	3		5	6	8	13	14	1	10	2		7	9	11	
Collingwood Harbour															
Severn Sound	3			6	8		14	1				7		11	
Spanish River Mouth	3		5?	6?		13?	14?	1	10			7			12
Lake Erie															
Clinton River	3	4		6	8	13	14	1	10			7		11	
River Rouge	3	4		6	8		14	1	10			7		11	
River Raisin				6				1				7			
Maumee River	3	4		6	8			1	10			7	9	11	
Black River	3	4	5	6	8	13?	14	1	10	2?		7		11	
Cuyahoga River	3	4	5?	6	8	13?	14	1	10	2?		7	9?	11	
Ashtabula River	3	4		6			14	1				7			
Presque Isle Bay		4		6?					10			7			
Wheatley Harbour		4?		6?	8?		14		10?			7			
Lake Ontario															
Buffalo River	3?	4	5?	6			14	1		2?		7			
Eighteenmile Creek				6?			14?	1?				7?			
Rochester Embayment	3	4?	5	6	8	13	14	1	10	2?			9	11	12
Oswego River	3	4?	5?	6?	8	13?	14	1						11?	
Bay of Quinte	3	4?		6	8	13	14	1	10			7	9	11	
Port Hope												7			
Metro Toronto	3	4?	5?	6	8	13?	14	1	10			7		11	
Hamilton Harbour	3	4	5	6	8		14	1				7		11	

▼ Table 1.1

Area of Concern	Ecological health and reproduction						Habitat	Human health			Human use/welfare			
Connecting Channels														
St. Marys River	3	4		6	8		14	1	10		7		11	
St. Clair River		4?	5	6			14	1	10	2?	7	9	11	12
Detroit River		4		6			14	1	10		7	9	11	
Niagara River (ON)	3		5	6	8	13?	14	1	10		7	9	11	
Niagara River (NY)	3?	4	5?	6			14	1			7			
St. Lawrence River (Cornwall)	3	4	5	6	8	13?	14	1	10	2?	7	9	11	12
St. Lawrence River (Massena)	3?	4?	5?	6?		13?	14	1						

The numbers in this table identify specific use-impairment categories used in the Great Lakes Water Quality Agreement. (Question marks indicate the impairments being investigated.) The GLWQA lists 14 beneficial uses that may be impaired and in need of restoration. The four general categories below contain the 14 impairments identified by number based upon the sequence in which they appear in the agreement.

Ecological health and reproduction
Degradation of fish and wildlife populations (3)
Degradation of benthic populations (6)
Degradation of phytoplankton and zooplankton (13)
Undesirable algae/eutrophication (which may cause low dissolved oxygen levels that may in turn cause other impairments) (8)
Fish tumours and other deformities (4)
Bird or animal deformities or reproduction problems (5)

Human use (welfare)
Tainting of fish and wildlife flavour (2)
Restrictions on dredging (7)
Taste and odour in drinking water (9)
Degradation of aesthetics (11)
Added costs for agriculture or industry (12)

Fish and wildlife habitat (14)

Human health
Restrictions on fish and wildlife consumption (1)
Beach closings (bacteria) (10)

Source: Adapted from Environment Canada and US Environmental Protection Agency (1995), <http://www.epa.gov/grtlakes/atlas/use-impa.html>.

Hamilton, Ontario, and the major infrastructures of cities, such as the bulk of downtown Chicago, are located on the shores of the lakes.

Many of these uses impinge upon each other. "Negative inter-dependencies," as political economists term them, exist between many of the uses. Waste disposal can often hinder contact recreation; pleasure boating and marinas can destroy fish habitat; wetlands can grow and limit canoeing; canoeing can upset turtle populations. It goes on and on. In any relatively confined site on the lakes, too much of any single use (let alone multiple uses) can reduce the availability of that site for future use. You can fill up a wetland with building materials and other solid wastes, but the wetland itself is a finite dumping place. Thus there are not just negative interdependencies, but also what are called "common pools," where any use can subtract from future resource availability. Negative interdependencies (sometimes called externalities) and common pools (sometimes called common-property resources) are features of many major harbours, rivers, and estuaries.

Most communities strive to balance multiple uses in ways that fit their priorities. The environment may be balanced against commercial shipping uses – with some areas set aside for docks and piers, while others are reserved for wetlands – and the wastes treated to reduce the major contaminants. Some communities fail to balance uses adequately, and a use, or even a resource, may disappear completely. In the late nineteenth century, for example, the Great Lakes supported some major commercial fisheries based on the harvest of whitefish, trout, and salmon. By 1920, the whitefish had been fished out of Lake Superior, the salmon out of Lake Ontario, and the trout from all five lakes (Ashworth 1986, 117). In this case, the combination of common-pool characteristics (overfishing subtracted from the availability of the fish) and negative interdependencies (pollution from wastes) were not balanced with the value of commercial fisheries. Uses have been driven out all over the Great Lakes, but especially in AOCs. This has largely been inadvertent; however, occasionally it has been deliberate. Communities made little effort to devise rules to balance lake uses.

Rules and Institutions

Scholars have wrestled for over thirty-five years with common-pool and negative-interdependency issues. How can an appropriate balance of uses be sustained? Does the balance require new rules and institutions to police them? Can users figure out and implement their own rules? Should national and regional interests and their use priorities dominate local interests? What kinds of technical information are needed for wise management and governance?

No agreed-upon resolutions have emerged to answer such questions. We are probably better able to diagnose the environmental problems that need attention than we were thirty-five years ago; however, we are still not close to achieving general solutions.

One of the most promising lines of inquiry into common-pool resource problems, if not into the range of multiple-use interdependencies, is the "IAD" approach. IAD, or "institutional analysis and development," refers to the work pioneered by Elinor and Vincent Ostrom. Using concepts from economics, anthropology, law, and political science, IAD views all rules as incentive systems for individual and group behaviour rather than as government commands. Rules can promote negative as well as positive consequences, and the rules for the Great Lakes have featured both. Newer rules, like those in RAPs, add to the existing stock of international, federal, state/provincial, and local rules. We will look at the function and implementation of RAP rules within this wider setting.

The acronym "CPR" refers to common-property, or common-pool, resources. In the last decade, a major intellectual revolution has occurred in academic circles with regard to the analysis of CPRs. It used to be thought that CPRs were automatically subject to the "tragedy of the commons," where a resource would be exhausted, perhaps permanently, by users interested in their own short-run benefits. Examples of such "tragedies" included the North American buffalo, the North Sea cod, and Atlantic and Pacific killer whales. The task was to get authorities to limit withdrawals from, or access to, a common pool through legal or technological means.

This revolution in academic thinking came about due to mounting evidence that both governmental and private exclusion of users were failing to prevent tragedies, which were occurring because of ignorance or ineffectiveness on the part of both users and non-users. Moreover, there was increasing evidence that many smaller-scale communities were relatively successful in developing a sustainable resource over time and managing to avoid "tragedies." A combination of social norms and written laws, generically referred to as rules, could be fashioned by users for their own long-run benefits. Again, Elinor Ostrom was a pioneer in developing these studies, along with members of the International Association for the Study of Common Property.

What the existing state of knowledge lacks, however, are explanations of, and evidence concerning, how to govern and manage multiple-use interdependencies when some of the uses can be common pools. In other words, we have some understanding of incentive systems and their operations with regard to the analysis and development of public policies. We also have some understanding of the application and effectiveness of incentive systems with regard to the governance of single (and usually small-scale) resources. However, until now, we have had no analysis of the appropriateness of differing governance systems for managing larger-scale and interdependent uses of a common pool. The present study is designed to fill this gap. RAPs are thus intellectual as well as practical experiments.

Governments at Work
In a previous work (Sproule-Jones 1993), I focused upon the actual rules (called rules-in-use) rather than the paper rules (rules-in-form) used to govern and manage public policies in Canada. I then explained the basic logic pertaining to this second level of rules, which may be referred to as the "collective-choice level," where decisions were made to provide, produce, and regulate public policies. Finally, at a third level, the "operational level," I described the basic logic of the operation of three selected policies evidenced in one site in Canada. The purpose was to see how incentive systems were created, sustained, and changed as well as how governments really worked (rather than how they said they worked). I found a relatively stable set of constitutional rules, a wider set of collective-choice rules, and a plethora of operational rules that were tailored to different policies in different places.

Hamilton Harbour, on the shores of western Lake Ontario, was subjected to serious scrutiny in this earlier study. We will examine Hamilton Harbour again here, but only as one of the AOCs that developed a RAP. The harbour was and is a multiple-use site, with shipping, recreation, and wastes disposal as its major uses. In *Governments at Work* (Sproule-Jones 1993), I attempted to measure the performance of harbour agencies in managing each of the three uses and to disentangle the rules that jointly regulate the three multiple-use cases.

Because any major political system will have many constitutional, collective-choice, and operational rules for public policy, it becomes difficult to estimate the precise impact of any one rule or set of rules upon policy performance. We did note, however, that the rule of "navigable servitude," traceable to the Magna Carta, did seem to be the major operational rule-in-use in balancing multiple cases in the harbour. It gave priority to shipping and navigation. Other uses were driven out by this rule or by the technological interdependencies of other uses. So decades of infilling marshes and wetlands (over 20 percent of the harbour's open waters) severely reduced and eliminated the commercial fisheries. Waste disposal from municipal and industrial sources, untreated until the 1960s, eliminated all recreation except berths for pleasure boats. Struggles at the collective-choice level in the decision-making forums in Ottawa, Queens Park, Hamilton, and elsewhere have led to significant changes and improvements in the water quality environment. But the basic logic of navigable servitude persists. The basic logic of all three levels of rules seemed relatively stable and predictable, despite the constant threats to the political system from Quebec interests and from other centrifugal movements.

This study picks up from *Governments at Work* in some significant ways. It focuses upon grassroots democracy and the rules that enable groups of users of a common pool to cooperate and perform effectively. The remedial action planning efforts in question take place within a system of collective-choice rules and institutional rules in two countries and also internationally. They nest, to use a *Governments at Work* concept, and they are stacked (i.e., a different configuration of rules exists for each of the several multiple uses). Furthermore, the RAP efforts add to the nesting and stacking of rules by devising their own particular rules of decision making and implementation at the community level. In short, we look at local democracies of resource users but within a broad institutional framework.

We also compare AOCs. The *Governments at Work* study reflected, in part, the theoretical knowledge of the 1980s. New insights have been gained since then, and we borrow and extend these to look at the 43 AOCs. Some are highlighted in detail: Hamilton Harbour in Canada, the Menominee River in the United States, and the international waterways of the Niagara and St. Lawrence Rivers. Others are discussed in lesser detail (relying, in part, on a survey to gather information about their key features).

Finally, we attempt to offer some prescriptive solutions about better ways to construct local self-governance of natural resources. We compare systems to see whether there is a basic logic in some local RAP arrangements but not in others.

Theory and Practice

Many people involved in resource extraction or other uses of a common pool do not have the time and opportunity to engage in considered reflection with other users about their institutional arrangements. They may have immediate problems affecting their own interests that could perhaps be remedied by institutional changes. But they do not have the incentives to engage in institutional analysis and design of a more comprehensive kind.

Similarly, scientists working in government or university laboratories may not have the incentive systems to reflect on their roles and the effectiveness of the application of their findings. Knowledge for its own sake has considerable merit when one views the technological and theoretical achievements of physical sciences over the last few centuries. However, it may not be applicable to all situations, and it may not be responsive to situations that confront common-pool users.

In many situations, therefore, there can be an unfortunate and unnecessary disjunction between theory and practice. Institutional arrangements may get "pasted together" based on the experience and memories of resource users. And science may be both inapplicable for many resource problems and too costly to justify in a world riddled with competition for limited budgets.

It was in this kind of world that RAPs were formed and are being implemented. And, as a consequence of this, one should anticipate some errors. The test of the RAP process will be whether decision makers will learn from these errors or whether they will blame those factors (such as politicians) that are only tangentially related to the problems of the AOCs.

I became involved in the formation of the Hamilton Harbour RAP in 1986 as a participant representing McMaster University (a major landowner), with lands fronting on the tributary (called Cootes Paradise) to the harbour. In the jargon of sociological methods, I became both an observer participant (a stakeholder concerned with analysis of process) and a participant observer (an academic with access to a stock of knowledge pertaining to comparable experiences elsewhere). I later helped to draft the implementation plan for the Hamilton RAP.

Boundaries

The previous comments suggest that the boundaries between the academic enterprise (which entails reflection, analysis, and proof) and practical problem solving can contribute to self-governance on the part of communities of resource users. At a minimum, the academic enterprise can supplement the work of government scientists and consultants in searching for information pertinent to solving a perceived issue. More controversially, perhaps, the benefits of the relationships across this boundary are mutual. University scholars can learn, and develop generic expertise, from conducting research of an applied nature. There are two counterincentives to this mutuality. One comes from the agenda of some disciplines to focus on theoretical development at the expense of applied analysis. (The discipline of economics may be a case in point.)

The second counterincentive comes from the community, which may not be able to distinguish inadequate research from meritorious research. The community may also perceive analysis as functioning to marshall evidence to support preconceived results – a process similar to the consulting and legal methods of investigation – and it may demand that social scientists deliver this kind of analysis. Academic tenure is a valuable veto of behaviour that may be promoted by this last counterincentive. It had to be invoked in 1983 when one Hamilton politician demanded that I withdraw from harbour investigations. But there are few strategic instruments available to local communities seeking to involve academics in applied research across the "town-and-gown" boundary.

Boundaries can also exist, of course, between different community interests as well as between different university departments. The RAP process is a useful mechanism for surmounting difficult community interests. However, in addition, the direct involvement and support of chief administrative officers, executive directors, and other heads of organizations give legitimacy and effectiveness to the results of the process. Years of failure to overcome these community differences in the Hamilton case, particularly the differences between commercial and environmental interests and the development of the harbour, functioned to promote action towards, rather than despair over the possibility of, building a new community forum.

The boundary between university departments may be more intractable. A concern with immediate and short-term interests and a fear of investing resources in uncertain enterprises can lead many academics to conclude that the boundaries between disciplines have real academic, rather than merely administrative, merit. What damaged this attitude in our case was the success of a collaborative research application to fund multidisciplinary research on the RAP process in Hamilton. This funding, a $2.1 million grant from Canadian federal funding agencies (the Social Science and Humanities Research Council, the National Science and Engineering Research Council, and the Medical Research Council), was awarded for the three years beginning April 1993. The project, known as Ecowise, provided the RAP process with an important research arm at a time when governments were cutting environmental budgets at the federal, provincial, and municipal levels in Ontario. It also facilitated some significant intellectual boundary crossings.

Managing across Boundaries
Stakeholders can be individuals, representatives, or organizations

drawn from a variety of user groups; however, in the case of university-community relations and intra-university relations, they are coalition partners in prospective policy function and implementation. The key word is "partners" – a word that implies equal status. There is no single centre of power, just a community of understandings and of self-interest. This requires a management style and ethos of the type rarely taught in business schools and schools of public administration – a type that is rarely part of the working experience of organizational managers. Successful RAPs have developed the ability to manage across boundaries, particularly during the lengthy periods necessary for successful policy implementation, as have successful university presidents.

This management style has three necessary elements, the first of which is inclusivity. Policy formation and implementation requires the demonstrable support of as many actors as possible, partly for legitimacy, partly for funding (either directly or by advocacy), and partly for galvanizing recalcitrant but needed stakeholders through peer pressure. Naturally one gets free riders in such situations, but the emphasis on inclusivity poses the possibility of replacement by another party in a potential team effort by all the participants and is thus subject to social norms requiring participation by community leaders.

The second element, obviously related to the first, is team building. Positive incentives need to be pursued continuously. The media can play a particularly useful role in building team morale as they communicate common successes to wider audiences and reinforce the positive elements behind the social norms for participation. Both Ecowise and its RAP partners enjoyed considerable media success in demonstrating the practical values of academic research through the RAP process.

The third element is to deal with the high transaction costs of coalition decision making by placing as many of these as possible onto professional managers. This seems obvious, but experience suggests it is rarely practised. Organizations seem to develop into decision-making forums where the policy makers have to implement ideas, especially if they suggest and document their worth, and professional staff do the coalition building and policy scanning. The role reversal produces burnout and exodus on the part of stakeholders (particularly if they represent organizations) and bureaucratic free enterprise on the part of staff. Cross-boundary management needs to correct these tendencies if it is to endure over the long run. Universities can be inept at organizing faculty and staff to exploit their comparative advantages. Ecowise and the Bay Area Restoration Council (the Hamilton Harbour monitoring organization) were deliberately fashioned on different lines.

This Study

This study is part of a three-year effort by Ecowise, during which seventeen major projects were undertaken. This is the only project that took a comparative approach to analyzing the experience of the Great Lakes. The purpose of this was to gain insights into governance arrangements and performance.

Political scientists will find the study different from what they are accustomed to because it deals with the complex intersection of public and private actors in developing and implementing environmental policy. In its generalizations, the study does not treat the documents of Congress, Parliament, and other public bodies – or the insights of senior bureaucrats – as good evidence. It deals with policy as practised in the real world by multiple actors and stakeholders. Policies (e.g., environmental policies) are developed as human artifacts by intelligent citizens dealing with common problems. Vincent Ostrom has written most wisely on these fundamentals.

Economists will find some differences in this study as well. Institutional arrangements are taken as basic data rather than as exogenous to policy description and prescription. The theory and evidence developed in this book owes much to the institutional economics of John R. Commons, Frederick Hayek, and James M. Buchanan, who depict different markets evolving to deal with collective problems while being grounded in laws (like contract law) and understandings that make market operations feasible.

Physical and biological scientists will note how ecosystem terminology is used to describe social as well as biological interdependencies in a common pool. Complex interdependencies permeate living systems, including human systems, and their relations to the other biological worlds.

Policy makers in Ottawa and other capitals may perhaps learn about policy "in the field" and use the generalizations about institutional arrangements to focus carefully upon analysis and design. They may recognize how local differences can call for different institutions and that remedies and adaptabilities are essential if enterprises are to function well.

The general public living around the Great Lakes will perhaps benefit the most, particularly if the environment of this huge watershed is gradually restored. The environmental "hot spots," the AOCs, must be restored first. As governments are still framing Lakewide Management Plans (LaMPs), success will likely come through RAPs. Citizens will first recognize improvements locally.

Format

We will be examining restoration activities in the 43 AOCs and explaining the relative successes achieved at these differing sites. We will assess how rules (institutional arrangements) operate with different degrees of effectiveness and efficiency in restoration efforts. And we will take into account the biophysical complexities and the magnitude of the restoration requirements at each site.

Chapters 2 and 3 set the background for the analysis. Chapter 2 discusses the uses of the Great Lakes over time and how the lakes have increasingly become places for waste disposal – for junk, sewage, industrial wastes, contaminated soils, agricultural soils, and so on. Four other major uses of the Great Lakes are also reviewed: commercial shipping; fishing, both commercial and recreational; water supply on the land; and hydroelectric generation. The economic magnitude and importance of the waste disposal use becomes apparent in our discussion. Chapter 3 reviews the major laws of two countries, eight states, and one province as they deal with environmental and resource management issues such as RAP restorations. The rules underlying joint national efforts, like the rules of the IJC, are also addressed.

Chapters 4 and 5 look at how institutions operate to develop and implement RAPs. Chapter 4 reviews the explanations of common pools, or common property. As noted, the Great Lakes represent a large pool of interdependent uses, and AOCs are also common pools where interdependent actors use the resource for their own purposes. Within these contexts, the rules for environmental management become important. Thus the theory on institutions and rules for common pools, which we develop in this chapter, sets the overall agenda for our study of different RAP operations. Chapter 5 develops this theory further, this time by offering explanations of how institutions can be designed with different environmental consequences in mind, remembering that different sites will pose different problems for stakeholders to solve. In this chapter we discuss our methods for gathering and examining the evidence, noting that we gather some evidence for all 43 AOCs and some extra evidence from particularly complex sites.

Chapter 6 sets out the findings, and Chapter 7 reviews them in the light of what we expected, and did not expect, to discover. There were some surprises. In essence, what we discover is that institutional design is critical for the success of any RAP. Some RAPs are efficient and effective because stakeholders are empowered to seek operational solutions and because implementers have incentives to manage across organizations in the public and private sectors. Other RAPs are a disappointment at either the planning stage or the implementation stage, and a few are disasters. If RAPs fail, then it is generally due either (1) to the indifference of government departments of the environment towards institutional design or (2) to their decision to construct the design of RAP institutions to suit their own narrow interests rather than those of local stakeholders. The restoration of the Great Lakes is a mixture of good, modest, and poor performance from one RAP site to the next. The promises are still to be kept.

2
History of Key Uses of the Great Lakes

The Great Lakes inspire evocative language and hyperbolic metaphors. Samuel de Champlain called these massive fresh waters the "sweet seas." Two hundred years later, the American explorer Henry R. Schoolcraft wrote of the "riches of the soil and the natural beauty of the country" on the "bold shores" of the Saginaw Valley in the Territory of Michigan. In 1996, the historian Pierre Berton (1996, 18) would write lyrically:

> Each inland sea is like a small nation in its infinite variety. Each has its own character ... I see Superior ... as remorseless and masculine. Huron, with its thirty thousand islands, reminds me of a fussy maiden aunt. Michigan, half wild to the north, heavily industrialized to the south, is an errant uncle. Erie is a wilful ingenue of changeable mood and false promise. Ontario is a complacent child.

The attitudes, values, and behaviours of settlers, both Aboriginal and European, often reflect the utilitarian rather than the lyrical values of these lakes and their resources. Six thousand years ago, the early inhabitants were hunters and gatherers who used the indigenous copper in their tools and weapons. Four hundred years ago, the European explorers sought a Northwest Passage to the riches of the Orient but returned with evidence of furs, fish, and trees for exploitation. And only a decade ago, a "Grand Canal Plan" to divert 30 percent of the discharge of the Great Lakes to the American southwest and drier regions of western Canada merited serious consideration by the then (Canadian) Inquiry on Federal Water Policy (Pearse, Bertrand, and MacLaren 1985, 126-29). The history of the Great Lakes can be seen as a history of the uses of the resources in the basin as well as a history of human stresses on its ecosystems.

The common law and customs of the early settlers reflected this utilitarian premise. English common law was structured along use lines. Different uses of natural resources required different legal solutions to conflicts that would arise over these uses. So the common law on fisheries, for example, differed from the common law on navigation and water transport. And when they came into conflict, the common law offered guidelines as to which use merited priority. For example, in both Canada and the United States, the common law rule of "navigable servitude" permits the uses of shipping and navigation to trump all other uses. This rule originated in the Magna Carta (Sproule-Jones 1993).

The restoration of the Great Lakes must take into account the many uses of the basin by its 33 million inhabitants and the demands on its waters by people living beyond its borders. These other values, and the rules developed for their "equitable" expression, provide the context within which policies for environmental restoration must interact. Some history of these alternative uses is thus in order.

Transportation

Human settlements and the exploration of the interior of the New World were intimately related to the transportation links provided by rivers and lakes. The St. Lawrence River and the interconnecting five Great Lakes provided an early and major example of this, at least for incipient European populations. The goal, in part, was the extraction of furs (particularly beaver). The means were largely canoes, which would use waterways and portages. The partners of the trappers and voyageurs were French missionaries, literate men with the skill to make maps and chart the miles of navigable waters (Kaufman 1989). Eventually, albeit fitfully, small enclaves of farmers and military families dotted the harsh and dangerous shores from the south shore of the St. Lawrence through to the Grand Portage on the far western shore of Superior in what is now the State of Minnesota (Ashworth 1986; Berton 1996).

The late eighteenth century saw the introduction of newer ships and ship technologies on the lakes. Some were naval vessels belonging to both sides in the American Revolution and the War of 1812. Some were merchant schooners built to ply their trade in goods and people from the Atlantic shores. Both were restricted by physical obstacles that connected the lakes. Canals were the solution in the St. Lawrence River, in the portages between Lakes Ontario and Erie, and between Lakes Superior and Huron. The building of the Erie Canal (1825), connecting Albany and Buffalo in New York (and, through them, Chicago and New York City), was a key precipitating factor in the construction of additional

Figure 2.1

The Great Lakes basin showing transportation links

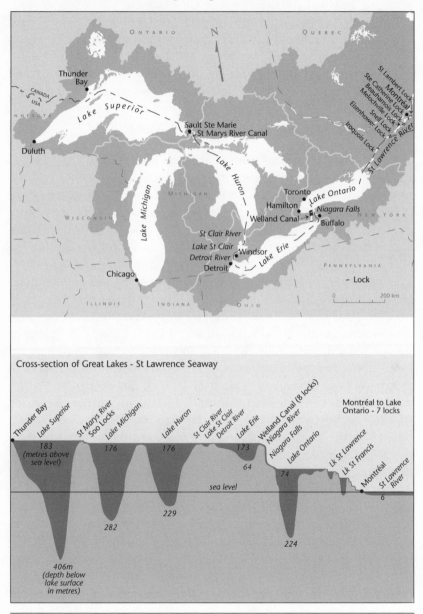

Source: Environment Canada, *Our Great Lakes, 1999*, <www.ec.on.gc.ca/glimr.maps-e.html>.

and deeper canals on the St. Lawrence and the first Welland Canal in 1829. It was not until 1855 that the "Soo" was circumnavigated by the state lock linking Lake Superior with Lake Huron.

The changes presaged a massive expansion of settlement and resources exploitation throughout the lower basin. Millions of European settlers moved across the lakes on steamships, which then returned with freight, particularly lumber and fish. Sailing ships continued to ply their trade; over 1,100 were registered with the Dominion government in the 1850s. Small human settlements burst with populations. Chicago, for example, went from 500 persons in 1838 to 2 million in 1900 (Ashworth 1986, 95-96).

Bulk cargo ships made from iron gradually replaced the wooden schooners. These were needed, inter alia, to transport iron ore to the new steel cities like Cleveland, Buffalo, and Hamilton. Vessel sizes went from 100 to 500 feet, and cargo weights from 400 to 4,000 tons. Shipping channels, especially the upgraded Welland Canals bypassing Niagara Falls, were as important as were railroads in transforming the Great Lakes.

Throughout the lakes, water transportation improvements in terms of canals, locks, and diversions were accompanied by port development. Hundreds of small harbours were developed, often by private companies, for trade in bulk commodities like iron ore and wheat. The prospect of direct competition with railroads over shipping bulk goods to Europe led shippers, shipping lines, and port managers to lobby for both capital improvements and institutional changes on the lakes. A fourth Welland Canal, with a draft of 15 feet, was opened in 1887; a new canal was constructed at Cornwall, and improvements were made to widen and deepen existing channels (Sussman 1979).

In 1871, the Treaty of Washington finally settled the exact international boundary between the two nations on the open waters, and, more important, the rules of navigation on the St. Lawrence River. Binational water issues, including the funding of commonly shared benefits of capital improvements, led to the creation of an International Waterways Commission in 1895 and, eventually, to the Boundary Waters Treaty of 1909. This treaty established the International Joint Commission (described in the next chapter), which had the power to examine and recommend solutions to binational problems along the international borders. Article 1 of the Boundary Waters Treaty is significant for our purposes. It states that the navigation of all navigable boundary waters shall forever continue free and open for the purposes of commerce to the inhabitants and to the ships, vessels, and boats of both countries equally, subject to non-discriminatory laws and regulations not incon-sistent with the privilege of free navigation. This article reconfirms the domestic policy of both countries and their incorporation of the com-mon law principle of "navigable servitude" into rules pertaining to shared international waters (Sproule-Jones 1993).

Later articles spell out some of the further implications for multiple water uses. In general, the IJC must approve new (post-1909) uses, di-versions, and obstructions of the flow and levels of waters on the Great Lakes. Article 8 attempts to prioritize uses in the approval process. Wa-ter supply and waste disposal within waters are accorded a higher prior-ity than is navigation, which itself has a higher priority than do power generation and irrigation. The act is silent, however, on ecosystem, rec-reational, and other uses. Nevertheless, the IJC has been involved in granting certificates of approval for specific building projects that could negatively affect navigation (e.g., the St. Lawrence Power Project, which regulates outflows from Lake Ontario) or positively affect it (e.g., the Toussant Causeway in the St. Lawrence, which reduces currents for ship-ping) (Chandler and Vechsler 1992).

The shipping and railroad interests became major competitors in the movement of bulk goods, both dry and liquid, during and after the two world wars. The shipping industry was particularly well positioned to transport iron ore pellets from Quebec and Labrador to both US and Canadian steel manufacturers on the Great Lakes. Grain could be moved downstream when the ships returned and thus feed European populations. Plans for a uniform draft of 25 feet to be built jointly by both countries under 1932 and 1941 treaties was defeated in the US Senate (under the two-thirds majority rule for treaty ratification). Some authors attribute this defeat to the power of US railroad interests (Toro and Dowd 1961; Sussman 1979), although the treaty included diver-sions for hydroelectric power at Niagara, construction to generate power at the St. Lawrence, as well as navigation concerns.

In 1951, the St. Lawrence Seaway Act was passed by the Canadian Parliament. The act authorized the construction of navigation works on the Canadian side of the St. Lawrence River from Montreal to Lake On-tario. Tolls were designed to pay for the project. The Americans took coordinated action on their side in 1954, building facilities to circum-vent the international rapids section. In 1959 the St. Lawrence Seaway was opened, providing a 27-foot draft from Montreal to Lake Erie. Navi-gation improvements were coordinated with hydroelectric construc-tion projects determined by the Hydro Electric Power Commission of

Ontario and the Power Authority of the State of New York. A St. Lawrence Seaway Authority (Canadian) manages five locks in the Montreal-Lake Ontario section plus the Welland Canal. The Seaway Development Corporation (American) manages two locks at Massena plus other American sections of the St. Lawrence. Despite these continual improvements, the seaway is still too small for the massive ocean container ships that typically have at least a 35-foot draft and exceed the maximum length (700 feet) of the locks (McCalla 1994, 172).

Consequently, Montreal has become (with the New York–New Jersey complex) a hub port with spokes for further re-transport of containers by smaller ship or by rail and road. The seaway relies on bulk cargoes, 70 percent of which are exportable grains. With the development of subsidized grain production by the European Union (EU), the seaway has seen its cargo flows decrease and its financial situation move from serious to perilous. Cargoes fell by 30 percent from the early 1970s to the early 1990s, and the number of commercial vessels fell by over 50 percent. The operating deficit rose from $2.1 million to $7.8 million between 1971 and 1991 (St. Lawrence Seaway Authority 1971, 1992).

For 400 years, the navigational and shipping use of the Great Lakes–St. Lawrence has enjoyed primacy, but now other uses are attaining relatively greater significance. Nevertheless, the ecosystem still remains substantially altered.

Hydroelectricity
Public works in the Great Lakes, and specifically in the Niagara–St. Lawrence Rivers, provided joint benefits (as well as costs) from the generation of electricity. Indeed, many of the construction projects for the seaway were to be constructed for the joint purposes of navigation and power (McCalla 1994, 159). Niagara power became instrumental in the industrial development of southern Ontario and western New York. The first US generating station at Niagara was opened in 1881, the first Canadian one in 1893.

Subsequently, Niagara hydroelectric power became a major issue in Ontario politics (Nelles 1974). Originally, a private monopoly was granted exclusive rights to generate power at Niagara (1892), but it was succeeded by a private cartel (1903), which was succeeded by a public utility that undercut private competition and became a major Crown corporation (1920s).

Today, Ontario Power Generation (formerly most of Ontario Hydro) is the second largest supplier of energy in North America (Vining 1981), with approximately 20 percent of its power originating in Niagara

Peninsula hydro (Colborn 1990, 58). Less than 10 percent of the power of the eight Great Lakes states is generated from water. Ontario Power Generation remains a key component of industrial development in Ontario and a primary user of the water of the Niagara River.

The non-Niagara hydroelectric systems are generally small in scale, if not in numbers. In Michigan, there are 113 plants, but they produce only 1.5 percent of the power needed to meet existing demands. There are 120 comparable plants in Wisconsin and similar facilities in New York State. The St. Marys River Dam and the Moses-Sanders Dam on the St. Lawrence are major plants. There are no estimates of the impact of hydro power on fish transportation and habitat, although the remedial action plan for the Nipigon River attempts to manage water levels to the advantage of both hydro and fisheries interests (SOLEC 1996a, 23).

Fisheries
The Great Lakes provided an abundant source of food for Aboriginal communities as well as a sustainable trade item (Doherty 1990). European settlers discovered more than 150 species of fish, and the abundance of fish was commonly noted in the accounts of early settlers (Wallace 1945). One 1813 account describes the whitefish population at Pasque Isle as so plentiful that "one might fill a net by simply casting it into the water from the beach" (Prothero 1973, 11).

As early as 1795 a commercial fishery was established on Lake Erie, and by the 1830s another major fishery was established on Lake Superior (Ashworth 1986, 116). By the turn of the century, Lake Superior was producing 8 million pounds of whitefish alone; Lake Michigan's trout fishery exceeded 3 million pounds; Lake Huron produced 7 million pounds of trout; Lake Erie produced over 33 million pounds in its commercial fishery; and Lake Ontario provided over 2 million pounds of whitefish (Ashworth 1986, 110-17; Wallace 1945).

It was, however, an open access common pool, with few limits on fishing. At particular sites, like Hamilton Harbour, voluntary restrictions were adopted by fishers to avoid overfishing and the potential exhaustion of stocks (Holmes and Whillans 1984). However, the fast transportation of fish to large markets was facilitated by railroad expansion, and overfishing and the destruction of habitat by urbanization patterns were beginning to severely restrict fish yields. By 1910, for example, the whitefish population in Lake Superior had been reduced to below commercial viability.

In 1868 the first Fisheries Act was passed by the Dominion government. This act was administered by the Department of Marine and Fisheries

(Pearse, Bertrand, and MacLaren 1985), and its main purpose was to control overfishing and to maintain stocks. It enabled the Dominion government to place restrictions on the taking of any fish species that showed signs of becoming exhausted. The Dominion could control fishing seasons and methods, and restrict the dumping of such materials as sawdust into rivers and creeks (Wallace 1945). Provinces could also enforce their own regulations in addition to those contained in the Fisheries Act (Pearse, Bertrand, and MacLaren 1985).

The American federal government was much slower to respond to the overfishing problem than was the Canadian federal government. Complaints from Canadian fishers about their American counterparts were recorded as early as 1872 (Wallace 1945). Unrestricted fishing on the American shore prompted the Department of Marine and Fisheries to lobby American federal and state authorities to act on the overfishing problem. Although many American representatives favoured such restrictions, the federal government had difficulty securing the approval of the state legislatures concerned (Wallace 1945).

Canada's fishery officials were involved in restocking programs and hatcheries. After 1875, restocking lakes with desirable sport fish was an accepted Canadian government practice (Wallace 1945). From this period on, Ontario maintained its own hatchery system. Records from 1853 indicate that the first successful hatching of fish eggs was conducted by Theodotus Garlick of Cleveland, who used eggs he had gathered at Port Stanley (Prothero 1973). By 1917, Canada had developed 61 hatcheries (Prothero 1973, 199), which stocked whitefish, salmon trout, salmon, speckled trout, and pickerel. For many years, those in the fishing industry believed that hatcheries were the solution to overfishing and pollution problems. By the 1930s, however, it became evident that no restocking program could alleviate the declining fish populations.

In 1908, with the Inland Fisheries Agreement, the two national governments made a modest attempt at establishing coordinated fisheries policies. This agreement consisted largely of enforcing poaching regulations in each other's waters rather than developing sustainable fisheries. In the 1920s, the sea lamprey invasions of the lakes above Ontario exacerbated the overfishing problem. Commercial fisheries catches were roughly stable at approximately 1 billion pounds from 1921 through 1981, until concerns about contamination of fish by toxic chemicals and heavy metals led to major reductions in allowable catches (Great Lakes Fisheries Commission 1979, 2000).

Since the 1950s, with the establishment of the Great Lakes Fisheries Commission in 1956 (subsequent to the Convention on Great Lakes

Fisheries that was ratified by Canada and the United States in 1955), there has been a joint attempt to develop sustainable commercial fisheries. The commission recommends measures that would permit the maximum sustained productivity of fish of common concern, and itself formulates and implements a sea lamprey control program (Great Lakes Fisheries Commission 1993). Its effectiveness depends upon the policies adopted and implemented by the respective federal, provincial, state, and Aboriginal natural resources agencies, each of which has one representative on five lake committees. The policies in question, however, seem to be dominated by the interests of sports fisheries.

Today, some 17 million "angler days" occur in four Great Lakes (excluding Ontario), with 12 million of these in Lakes Erie and Michigan. (An angler day is an individual who fishes for at least 20 minutes during any one day.) There are over 6 million angler days on Lake Ontario (Canada [DFO] 1999), and the economic value of the sport fishery is estimated at over US$2 billion (Talheim 1988, 11). In contrast, the major commercial fishery, based in Lake Erie, consists of only 750 fishers who land 50 million pounds per annum (United States EPA and the Government of Canada 1995, 20) and the estimated landed value for all five Great Lakes is only US$40 million (Talheim 1988, 25). The United States prohibits the sale of fish affected by toxic contaminants, and fish consumption advisories (recommended maximum individual annual consumptions of different species) exist for all lakes. It is the sport and thrill of the catch that remains the most valuable of the attributes of a once massive natural resource.

Water Supply

Water is withdrawn in large volumes from the Great Lakes – 56,920 million gallons per day, or 2,493 cubic metres per second (IJC 2000, 7) – and then put to a variety of uses on the land. About 5 percent of these volumes is actually withdrawn and consumed; the remainder returns as flows from "on-the-land" uses.

The IJC estimates that 29 percent of the withdrawals are used for irrigation, 28 percent for public water supply, and 24 percent for industrial uses. Of the remainder, 6 percent goes for hydroelectric and thermoelectric power generation, 4 percent for self-supplied domestic uses, and 3 percent for livestock watering. The volume of groundwater withdrawals, as opposed to surface water withdrawals, is unknown (IJC 2000, 8).

Water is also removed from the basin by diversions. The largest diversion, 91 cubic metres per second, occurs in Chicago (IJC 2000, 10). A canal was built in the mid-1800s to divert water from Lake Michigan into the Des Plaines River, which empties into the Illinois River (which is part of the Mississippi River basin). The original reason for the canal was to allow transportation between the Great Lakes and the Mississippi; however, later it was used for public water supply and sewage disposal. The Chicago diversion is actually an out-of-basin diversion and is more than compensated for by an into-the-basin diversion at Long Lac and Ogoki on Lake Superior. There, 158 cubic metres per second are diverted for hydroelectric power purposes by the province of Ontario.

Other smaller diversions occur on the lower Great Lakes, and removals from the basin occur in the forms of bottled water, slurry, and ballast water. It appears as if the basin imports fourteen times more water than it exports (IJC 2000, 11), but the prospect of tanker withdrawals in 1997 and 1998 led to major political controversies throughout the region. Major new diversions or removals from the basin would violate the Great Lakes Charter of eight US states and the provinces of Ontario and Quebec. The charter requires the consent and concurrence of other jurisdictions should withdrawals exceed 5 million gallons/day in any 30-day period. Further, domestic US and Canadian legislation would require approvals for water removals from the basin. These legal arrangements appear to have done little to assuage the emotional concerns about water exports, especially in Canada.

Waste Disposal

The Great Lakes basin experienced massive population increases during the nineteenth century. Over 10 million people had settled on the shores by 1900, with the large majority of these being in the Lake Michigan and Lake Erie subsystems (Environmental Atlas and Resource Book 1995, 18). Urbanization and industrialization occurred, with major population nodes around southern Lake Michigan, western and southern Lake Erie, and the northwest shores of Lake Ontario. Heavy manufacturing (e.g., steel, paper, and chemicals) provided the primary sources of urban employment, while water and electrical power provided cheap sources of energy. Chicago proved to be a "model": it went from 500 people to over 2 million in the 70 years preceding 1900. Chicago was the confluence of agricultural and metals processing that expanded in tandem with the cheaper rail and water transportation routes to the east.

The expanded human settlements were largely cavalier in their disposal of wastes. Many cities like Chicago and Hamilton experienced cholera epidemics due to contaminated well and nearshore water intakes (Sproule-Jones 1993, Chap. 7). Solid wastes were frequently dumped into marshlands, which were considered non-valuable sites that could not support buildings and roads. Hamilton Harbour, for example, was eventually reduced by 25 percent, and this experience was replicated throughout the urbanized portion of the Great Lakes. Industrial wastes were either buried or diluted and dumped into nearshore waters and wetlands. Some communities were even proud of their air pollution, believing it to be a sign of progress.

Today, 33 million people live in the basin, and 80 percent of these live in 17 metropolitan areas (11 in the United States and 6 in Canada) (SOLEC 1996b, 5). The Conservation Foundation estimates that each resident generates about 2 kilograms (4.4 pounds) of solid wastes per year (Colborn 1990, 67). Most of these kinds of wastes are placed into landfills rather than dumped into rivers or lakes. However, rainfalls and snowmelts exceed evaporation throughout the basin, and, consequently, the leachates are likely to enter some parts of the water courses. In addition to these dumps, some 116 sites on the US side of the basin are considered hazardous (and part of the US's former Superfund program), and roughly 5 percent of 342 Ontario dumps are comparable (Colborn 1990, 63). Liquid wastes are dumped directly into the lakes and streams. These are estimated, albeit widely, at some 57 million tonnes per year (Colborn 1990, 64).

Wastes also enter the ecosystem through soil erosion, pesticide use, and manure "management" practices on agricultural land. Estimates of the scale of these activities are available, but there are no estimates of the extent to which they end up in non-point source pollution (SOLEC 1996b, 10). Similarly, wastes washed into storm drains, creeks, rivers, and bays from non-agricultural lands may be highly polluted, but again, no quantity estimates are available.

The impacts of these waste disposal activities on the environment intermingle with the impacts of direct human alterations of the ecosystems. Human settlements alter habitats through land development and/or resource extraction, while they alter hydrology through diversions and dredging. Increased sedimentation and sediment transport alter the physical processes on (at least) the nearshores, and biological structures are altered through human disruptions of an ecosystem's food webs (SOLEC 1996c, 39-44). The exact contributions of each of these is impossible to estimate.

The overall consequences show up in degraded water quality conditions and impaired natural ecosystems. Some of these consequences are well known. In 1953, the bottom waters of Lake Erie showed the first signs of anoxia. By the 1960s, the lake was often characterized as "dead," which meant that massive algal blooms were occurring and that several nearshore areas were largely devoid of aquatic life (Colborn 1990, 95). Lake Erie was subject to "cultural eutrophication," whereby phosphorous (as a nutrient) was imposing an algal bloom in a relatively shallow lake, which, in summer months, has relatively low dissolved oxygen levels. Public concern about Lake Erie was, in large part, instrumental in establishing the Great Lakes Water Quality Agreement between the United States and Canada (see Chapter 3). The practical consequence of the agreement was the reduction of phosphorous loadings through improved sewage treatment. Targets and objectives were attained by 1991, and chlorophyll *a* (an indicator of nuisance algal growth) was at acceptable levels by the early 1990s. Conversely, the introduction of the zebra mussel has increased water clarity (77 percent between 1988 and 1991) because of its filtration activities, and aquatic plants are thus spreading into deeper waters (Environment Canada/US EPA 1995).

In 1976, President Carter declared the Love Canal (near the Niagara River in New York State) a federal disaster area. The Hooker Chemical Company had buried 20,000 tonnes of chemical waste that, 30 years later, was seeping into water and air. The effects were showing up as human skin rashes and eye irritations, and detectable levels of persistent toxic chemicals in downstream fish tissue (Levine 1982; Colborn et al. 1990, 56-60). The leachate from Love Canal is now contained and treated, but the Niagara River is still the most significant source of toxic chemicals entering Lake Ontario. Upstream and downstream monitoring indicates that point and non-point sources on the river itself are significant modes of entry for these toxins (Environment Canada et al. 1995).

These two well-known cases of environmental degradation do not amount to an accurate story of the Great Lakes environmental situation. In Chapter 1, we noted the 14 indicators of impaired beneficial uses that need to be restored, at least in part, for the 43 AOCs that were designated by the IJC.

A joint publication of both national governments, *State of the Great Lakes 1995* (Environment Canada/US Environmental Protection Agency 1995) provides some summary overview indicators that extend beyond AOCs. The study suggests that four sets of indicators provide useful reviews of all the lakes. These indicators are:

(1) *Aquatic habitat and wetlands.* There are major losses throughout the basin: Ontario has lost 80 percent and the other lakes some 60 percent since the 1780s. Enhancement and restoration programs

cannot keep up with current habitat losses, with perhaps the exception of the brook trout habitat in the upper lakes.

(2) *Persistent toxic substances.* Loadings of persistent toxic contaminants have been reduced substantially since 1970, and there are declining contaminant concentrations in waters, sediments, fish, and wildlife. In urban areas and for certain species, however, levels are high enough to cause concern. For example, human consumption of top predator and forage fish is discouraged, and there are observed effects of alteration of biochemical function, pathological abnormalities, tumours, and reproductive abnormalities in well-studied species like the herring gull. Table 2.1 lists the critical pollutants that have been referenced by regulatory agencies.

Table 2.1

List of critical pollutants referenced by regulatory agencies

Chemical	GLWQA Annex 1	GLWQI	LaMPs critical pollutants	Pollution prevention	IJC list of 11 critical pollutants	Lake Superior priority substances	COA Tier 1 list	COA Tier II list
Aldrin	√				√		√	
Benzo(a)pyrene	√		√		√		√	
Chlordane	√	√	√			√	√	
Copper	√	√	√					
DDT and metabolites	√	√	√		√	√	√	
Dieldrin	√	√	√			√	√	
Furan	√		√		√		√	
Heptachlor	√	√	√					
Heptachlor epoxide	√		√					
Hexachlorobenzene	√	√	√		√	√	√	
Alkylated lead	√		√	√	√		√	
α Hexachlorocyclohexane	√		√					√
β Hexachlorocyclohexane	√		√					√
Mercury	√	√	√	√	√	√	√	
Mirex	√		√		√		√	
Octachlorostyrene	√		√			√	√	
PCBs	√	√	√		√	√	√	
2,3,7,8-TCDD (a dioxin)	√	√	√		√	√	√	
Toxaphene	√	√	√		√	√	√	

Abbreviations: GLWQA = Great Lakes Water Quality Agreement GLWQI = Great Lakes Water Quality Initiative LaMP = Lakewide Management Plan IJC = International Joint Commission COA = Canada-Ontario Agreement

Many persistent toxics reside in sediments and represent pollution loadings from previous years and generations. Thirty-eight of the 43 AOCs have restrictions on dredging because of the potential impacts of toxic pollutants in sediments. In-place contaminated sediments can have direct impacts on aquatic life, such as fish, and these may get redistributed through the food chains. Indirect impacts can occur when sediments are resuspended (through ship movements) or dredged and dumped in confined disposal facilities or shallow waters and on beaches. Unless in-place treatment of contaminated sediments is possible, then sediments must normally be dredged and then coupled with treatment technologies – biological, chemical, or thermal. Typical contaminants in the sediments of AOCs are heavy metals, PAHs, and PCBs caused by industrial discharges as well as persistent pesticides and herbicides from non-point sources.

(3) *Eutrophication.* Loadings of total phosphorus are at good or restored levels under the Water Quality Agreement targets. These objectives were also achieved by 1990 for total phosphorus concentrations in open water. Outside some 21 AOCs, levels of chlorophyll *a* are also considered as good or restored. Levels of dissolved oxygen, except for Lake Erie's central basin in summer months, are similarly restored. Nitrate-plus-nitrate levels for the open lakes show some increases, however, especially for Ontario.

(4) *Human health effects.* Thirty-five of the AOCs have fish consumption advisories, as noted above. The impact of toxics in general on humans is both complex and uncertain. Several recent studies associate increased tissue levels of toxic substances with reproductive, developmental, neurological, endocrinological, and immunological problems (SOLEC 1986a, 107). And there are studies linking PCB levels (and some other organochlorides) in blood plasma with fish eating. Ontario and the Great Lakes states have had fish consumption advisories (recommended portions per annum) since the 1970s. Twenty-four AOCs have beach closures or recreational body contact restrictions due to pathogenic pollution. Forty-four percent of the Canadian Great Lakes beaches are closed for at least some of the three-month summer (June to August), and 23 percent of US beaches are also closed during the same time period. (Monitoring of beaches varies by jurisdiction, as do criteria for closures.) There is more recent evidence and concern that the use of chlorine as a major cheap disinfectant in water supply systems from the lakes (as well

as sewage inputs) is associated with bladder and colon cancers. The dose-response linkages remain unclear, however (SOLEC 1996a, 103-5).

This suite of indicators provides a cursory overview of the pollutant impacts of waste disposal from humans and their activities throughout the Great Lakes basin. Particular sites, like AOCs, can differ. This fact becomes important when analyzing the effectiveness of restoration attempts. The institutional responses to these waste disposal and pollution issues form the basis of the next chapter.

Conclusion

Five major human uses of the Great Lakes have been described. Perhaps the most important one, certainly for the purposes of assessing restoration efforts, is that of waste disposal. Human settlements and the extraction and processing of natural resources in industrial sites on the Great Lakes have generated, and continue to generate, residuals that are either dumped directly or filtered indirectly into the ecosystem. A suite of indicators of their impacts is suggested, including impacts on living systems like fish, wildlife, and mammals.

Humans have also significantly altered the shorelines and connecting channels of the lakes in order to provide shipping channels and hydroelectric power. These, in turn, have made possible a commercial and, especially, a major recreational fishing and boating industry. The interdependencies between uses are also apparent, and we must now examine the framework of rules that has been devised by the major governments around the basin in order to deal with them.

3
Institutions and Rules for the Environment of the Great Lakes

Remedial action plans and the rules by which they are formulated and implemented "nest" deep within a large corpus of American, Canadian, and international law and practice. They are the product of statutes, regulations, and the common laws of the US federal government, eight US states, the Canadian government, the Province of Ontario, and international bodies – especially the International Joint Commission. The rules intersect and support RAPs for the 43 Areas of Concern on the Great Lakes. It is in this sense that RAPs "nest."

RAPs are designed with an ecosystem principle in mind. Essentially, this means that the interdependencies between users and uses of the basin ecosystem must be acknowledged and included in planning objectives and actions. Then the cross-media implications of impaired beneficial uses of the AOCs waters also have to be acknowledged; that is, the interconnections between air and land and the waters with which they share the environment must be included in the plans. The institutional implications are that the pertinent rules devised for use of land and air resources also have relevance for RAPs, as do, of course, the rules for water resources.

Previous studies of environmental values tend to neglect cross-media interdependencies (Johns, 2000). The early studies in Canada and the US focused on point source regulations and the rules of national governments (Caldwell 1963; Dwivedi 1974, 1980; Davies 1970; Sproule-Jones 1981; Marcus 1980). Later studies focused on cross-media pollution problems and their resolution in more comprehensive national legislation like the 1980 Comprehensive Environmental Response in the United States and the 1988 Canadian Environmental Protection Act (Ringquist 1993). Gradually, a body of knowledge about subnational governments, the role of federal arrangements, and, increasingly, the role of organized interests and associations has been sketched (Doern and Conway

1994; Filyk and Coté 1992; Wilson 1992; Rosenbaum 1991; Ingram and Mann 1989; Fiorino 1995). Comparative studies reveal that many developed countries define their environmental problems in the same way and use similar kinds of statutes and other rules (albeit dis- and reaggregated through the interactions of national and subnational governments within federal systems) (OECD 1996; Lester 1986; Hoberg 1992, 1993; Skogstad and Kopas 1992; Harrison, 1996).

This book, especially this and the next two chapters, focuses on localized sites within a large ecosystem, on AOCs within the Great Lakes ecosystem. These are simultaneously both subnational and international focuses, or units of analysis. We are letting the technical characteristics of an ecosystem and its highly stressed locations define a political-economic analysis of institutions. This is a radical departure from tradition, and it yields useful conclusions about institutional analysis and design that can be incorporated into public policy actions (Sproule-Jones 1993; V. Ostrom 1980, 1991; E. Ostrom 1990; E. Ostrom, Gardner, and Walker 1994).

Our review of the rules will start at the international level, with an examination of the International Joint Commission and how its pollution control policies evolved into ecosystem management policies. We will then describe the major components of US and Canadian legislative frameworks that must implement regulations and "manage" resource uses in the AOCs.

International Arrangements

Boundary Waters Treaty, 1909
The US and British governments recognized the importance of the Great Lakes immediately after the American Revolution. The peace treaty established the split jurisdiction over the lakes but retained exclusive British jurisdiction over the lower St. Lawrence. As the population and the uses of the Great Lakes grew in number and intensity, conflicts began to arise over "management" and priorities. National level disputes arose in the first decade of the twentieth century due to hydroelectric and recreational conflicts on the Niagara River (Royal Society 1985).

To help resolve such disputes, the International Waterways Commission was established in 1905. It was purely an investigatory and advisory body to the two national governments. It recognized the limitations of its powers and recommended the establishment of a body with the authority to make (limited) binding decisions on any use of boundary waters that affected levels and flows on either side. The result was the

1909 Boundary Waters Treaty and the establishment of the International Joint Commission (Muldoon 1983; Chandler and Vechsler 1992).

The IJC consists of six commissioners, three appointed by the president of the United States and three by the United Kingdom (after the 1929 Treaty of Westminster, by the government of Canada) on recommendation of the Canadian Cabinet (Article VII). However, in comparison with national governance regimes, the IJC remains relatively powerless. First, it must approve new uses, obstructions, or diversions of boundary waters if these affect the "natural" levels or flow of waters on the other side of the boundary (Articles III, IV, and VII). However, at least one government must refer such matters to the commission for consideration and approval, and the IJC has no power to enforce either the treaty or its own orders.

Second, the IJC may advise the government on matters referred to it for consideration (Article IX). For our purposes, Article VII of the 1978 Great Lakes Water Quality Agreement between Canada and the United States serves as a major example of the IJC's advisory role. The IJC was asked to assist in the implementation of the GLWQA by advising the national governments on progress towards the achievement of the agreement's objectives. We will examine this agreement and its implementation in more detail below.

Third, the IJC has an administrative role in relation to the two transboundary rivers – the St. Marys and the Milk Rivers. Under Article VI of the treaty, these two rivers are to be considered as one and the waters are to be shared equally between the two countries. In all its decisions, the commission is required to set priorities for water uses, placing domestic and sanitary purposes first, followed by navigation uses, and then power and irrigation uses (Article VIII). This third set of powers, albeit narrow in scope, gives the IJC directives some authority.

A fourth set of powers has never been used. Article X gives the IJC the legal power to arbitrate issues referred to it by both governments. However, neither government has seen fit to ask the IJC to arbitrate, preferring, instead, to turn to special conventions (Chandler and Vechsler 1992, 267).

The IJC exerts its powers through its reputation as an objective, nonpartisan authority that relies on the best scientific advice. It also gains power from its ability to mobilize large numbers of scientific experts (from both sides of the border) to staff its advisory committees. Rabe and Zimmerman (1995, 67-70) suggest that a policy community of environmental policy professionals with similar beliefs and value orientations, many of whom sit on IJC standing advisory and ad hoc study boards, provides for the substantial integration of environmental management across international and subnational boundaries. Nevertheless, the IJC's authority is largely non-legal and non-formal. Muldoon argues that the IJC has circumscribed its authority by limiting its scope to boundary waters and to various scientific and technical questions. This criticism, however, was written before the RAP program was introduced (Muldoon 1983, 25-6).

In sum, "the Commission has been created to serve the Canadian and the United States governments by assisting them in avoiding and resolving contentious or potentially difficult issues. The Commission fulfills this role by responding to References and by deciding on Applications for Approval to use, obstruct or divert boundary waters and transboundary rivers." In both cases, the two governments decide when a matter will come before the commission, and the commission is "dependent on the governments for [its] financial and personnel resources" (Chandler and Vecshler 1992, 267-68). Nevertheless, the IJC's prestige and reputation allow it to provide constructive criticisms and suggestions to governments regarding how to improve their water quality regimes.

Foundation of the Great Lakes Water Quality Agreements

As early as 1912, the Canadian and US governments asked the IJC to examine the general extent of pollution in the Great Lakes and to make specific recommendations for the St. Marys, St. Clair, Detroit, Niagara, and St. Lawrence Rivers (Hartig and Thomas 1988). The main concern of this investigation was the effect of waterborne diseases (such as cholera) on the burgeoning lakeside populations. The IJC recommended using sewage treatment and water purification to control human waste disposal. Waste disposal continued to be a problem until the eutrophication of Lake Erie captured public attention in the 1960s.

The Great Lakes Water Quality Agreement of 1972 came after several IJC studies on the Great Lakes. The focus on cleanup began in 1960 when the IJC appointed two joint investigative boards for Lake Erie, Lake Ontario, and the St. Lawrence River (Caldwell 1988). These boards were made up of technical people from federal, provincial, and state government agencies. Their interim reports drew extensively upon earlier scientific studies and emphasized nutrient loading (especially phosphorus) as the cause of eutrophication in the lakes.

The IJC forwarded a report to the two countries in the fall of 1965, recommending that measures be taken to limit phosphorus inputs. However, Ontario and the various American governments were not willing to consider such measures, and the report was set aside. More studies on the Great Lakes continued through 1966 and 1967 (Munton 1980, 155).

In order to study pollution in Lakes Erie and Ontario, the IJC established two technical advisory boards: the International Lake Erie Water Pollution Board and the International Lake Ontario–St. Lawrence Water Pollution Board (Muldoon 1983, 58). In 1969, a final report from these technical boards was presented to the IJC. It concluded that Lakes Erie and Ontario, as well as the international section of the St. Lawrence, were being polluted, especially by phosphates and other nutrients. This report was the first to include documentation that outlined the relative contribution of pollutants from both countries, and it made it clear that most pollutants originated from the United States (Munton 1980).

The first ministerial-level meeting devoted to the Great Lakes pollution problems took place in June 1970 (Muldoon 1980). The two sides agreed that transboundary pollution was contrary to the obligations set forth for each country under the Boundary Waters Treaty. However, the United States was not yet prepared to approve the IJC's recommendations to move towards taking remedial action in order to deal with the cleanup of the Great Lakes, nor was it prepared to begin negotiations. There was a significant bilateral disagreement concerning the fundamental premise underlying any abatement programs. Canada felt that, in effect, the 1909 treaty gave each side the right to contribute pollution that was up to 50 percent of the "assimilative capacity" of the waters. In order to meet this understanding, the United States would have to institute drastic pollution control measures, whereas Canada would have to institute only minimal cutbacks. Canadian diplomatic pressure on the US government ensued, and the task force report was submitted to the president. The report recommended that a series of positive steps should be taken to comply with the Boundary Waters Treaty. This resulted in a proposal to establish a joint working group that would examine the present programs and the possible need for a Great Lakes Water Agreement.

Both the Canadian and American national governments accepted the report of the Joint Working Group in June 1971, and final negotiations on a binational agreement began.

Great Lakes Water Quality Agreement, 1972

The 1972 GLWQA was designed to cultivate binational and intergovernmental cooperation with regard to addressing pollution issues in a large water basin. In this agreement, the IJC was responsible for collecting and analyzing information on objectives and programs for the verification of data that had been collected in the 1969 Lakes Erie and Ontario–St. Lawrence study. This information was to be analyzed by both a water quality board (which was to succeed the Connecting Channels Board) and a research advisory board, whose purpose was to coordinate further scientific work. Major studies were subsequently conducted on all the lakes. Many land-based activities that were believed to have had an impact on water quality were also studied (Colborn et al. 1990).

The document's primary focus, however, was the concern over excess nutrient loading. The main strategy used to reduce the nutrient inflow involved improving municipal sewage treatment. Between 1972 and 1978, about US$10 billion was spent on upgrades (Colborn et al. 1990). The effect on water quality was immediate and readily apparent. Over the period between 1968 and 1985, total phosphorus concentrations for Lake Erie declined at high detectable rates (GLWQB 1987, 84). Similarly, annual surveys from 1973 to 1986 indicated comparable rates of decrease in phosphorus loadings in Lake Ontario (GLWQB 1987, 84). Problems such as "vessel wastes, pollution of the Upper Lakes and pollution from land use (non-point sources) had proved too contentious or too complex" to warrant similar action (Munton 1980, 163). These sources were identified in the 1972 GLWQA but were only given to the IJC for further study. The 1972 agreement paved the way for the more comprehensive 1978 agreement.

Great Lakes Water Quality Agreement, 1978

The 1978 GLWQA shifted the emphasis away from excess nutrient loadings and towards control of toxic substances. It extended its mandate to cover all five lakes and their tributaries (rather than staying with Lakes Eire and Ontario), but it maintained the basic structure of the 1972 GLWQA. Joint water quality objectives and standards were set, commitments to implement control programs were made, and the IJC mandate of monitoring progress was revamped and continued. Additionally, more stringent standards were set across the board with regard to water quality. These measures included implementation of municipal and industrial abatement programs, reductions in nutrient loadings, and a decrease in toxic chemical discharge. And a new concept was introduced: the Great Lakes basin ecosystem. This term was defined as "the interacting components of air, land, water and living organisms, including man, within the drainage basin of the St. Lawrence River at or upstream from the point at which this river becomes the international boundary between Canada and the United States" (Royal Society 1985, 10).

This definition is officially used in conjunction with the GLWQAs. However, it is recognized that there are difficulties in attempting to define a term as broad as "ecosystem." Essentially, an ecosystem is a system of interacting components, both physical and biological, that exhibit some degree of internal linkage and an implied boundary (albeit an inconsistent and temporally indistinct boundary). Improving and sustaining the water quality of the Great Lakes is directly related to the linkages within the ecosystem and to how they are viewed both inside and outside the Great Lakes basin. However, there are continuing arguments regarding what should and should not be included within the confines of the Great Lakes ecosystem. The Royal Society of Canada states that this ecosystem can be simplified in terms of scale (Royal Society 1985). For example, the scale may be altered based upon whether one sees humans as external or internal to the ecosystem. Or it may be altered according to how one views the interactions between the lakes. Are exchanges between the lakes so minimal that each lake should be considered a separate ecosystem? According to the Royal Society's definition, human beings cannot be separated from the Great Lakes basin environment (which includes air, land, water, and other living organisms).

To alleviate such controversies, the 1978 GLWQA declares that the restoration and enhancement of boundary waters cannot be achieved independently of the restoration and enhancement of those parts of the Great Lakes basin ecosystem with which the waters interact. In light of this, an ecosystem approach to pollution would mean that government action affecting the lakes would proceed on the understanding that the bounded field of policy includes the basin-wide watershed of the Great Lakes and the multifarious relationships that affect it, both within and without (Caldwell 1988). Annex 17 of the 1978 GLWQA indicates that research and development should be conducted in order to

> determine the mass transfer of pollutants between the Great Lakes Basin Ecosystem components of water, sediments, air, land and biota, and the processes controlling the transfer of pollutants across the interfaces between these components (IJC 1989, 80-81).

As a result of an increased understanding of the dynamics of pollution in the Great Lakes system and the emergence of new concerns (such as toxins and the effects of non-point source loadings), the new GLWQA was signed in 1978. This document recognizes that water quality is dependent upon the interaction of the entire ecosystem: air, land, water, and the activities of living organisms (including human beings). The purpose of the 1978 GLWQA is to

> restore and maintain the chemical, physical and biological integrity of the waters of the Great Lakes Basin Ecosystem. In order to achieve this purpose, the Parties agree to make a maximum effort to develop programs, practices and technology necessary for a better understanding of the Great Lakes Basin Ecosystem and to eliminate or reduce to the maximum extent practicable the discharge of pollutants into the Great Lakes System (IJC 1989, 7).

The GLWQA may be viewed as a reaffirmation of the intent to prevent and control Great Lakes degradation. It is based on the 1972 agreement as well as on further revisions that have occurred in light of interim scientific research. Its general objectives detail qualitative standards for water quality – standards that strive for water free from such substances as floating materials, heat, and nutrients. It attempts to eliminate any substances that may be detrimental to the quality of the water; the utility of the water; or the health of the water's users, including human beings, waterfowl, and aquatic life (Bixby 1986).

The GLWQA's specific objectives specify the minimum levels of water quality desired. These levels are set forth in quantitative terms, such as concentrations of particular substances within the Great Lakes system. The objectives are based on available information of cause-and-effect relationships between pollutants and receptor points that will, in theory, protect most sensitive uses in all waters. They may be amended, or new objectives added, through the mutual consent of both parties. Water quality is classified into chemical, physical, microbiological, and radiological properties. The chemical properties section is further subdivided into persistent toxic substances, both organic (e.g., DDT) and inorganic (which includes metals, e.g., arsenic/lead). Non-persistent toxic substances include organic (e.g., oil/pesticides), inorganic (e.g., ammonia), and other substances (e.g., dissolved oxygen/pH/phosphorus). The physical properties section includes asbestos, water temperature, and suspended solids. The microbiological properties section states that "waters used for body contact recreation activities should be substantially free from bacteria, fungi or viruses ... [that may cause] human diseases or infections" (IJC 1989, 27). The radiological properties section identifies the limits of "safe" levels of radioactivity in water used for human consumption.

While the GLWQA has these provisions, it is not binding. It provides a framework within which the governments can manage Great Lakes water quality.

The most fundamental bilateral disagreement during the negotiation of the 1972 and 1978 GLWQAs centred around the so-called "equal rights" of each country to the use of the waters. If both countries are assigned equal rights, then each must control its waste disposal so that it contributes no more than half of the total pollution. If there are no equal rights, then both sides have to take equivalent measures to control pollution. Canada, which contributes substantially less pollution than does the US, advocated the former position, while the United States advocated the latter (Munton 1980). The debate is still unresolved. Both parties agreed, in the interim, to reduce pollution without compromising the two divergent principles.

The Great Lakes Charter, 1985
While the GLWQAs were being formulated, the states and provinces were providing their own "agreement" in the form of the Great Lakes Charter. This charter represents a unique bilateral effort to clean up the Great Lakes basin at the subnational level. In the late 1970s and early 1980s, the eight Great Lake states realized that the threat of diversion to arid parts of the American southwest was real. The region's governors and the premiers from Ontario and Quebec decided to adopt a new policy resolution. The Great Lakes Charter was signed in 1985 as an act of good faith, and its purpose was

> to conserve the levels and flows of the Great Lakes and their tributary and connecting waters; to protect and conserve the environmental balance of the Great Lakes Basin ecosystem to provide for cooperative programs and management of the Great Lakes Basin by the signatory States and Provinces ... and to provide a secure foundation for future investment and development within the region (in Bixby 1986, 23).

The purpose of the Great Lakes Charter is to provide the opportunity for basin-wide management. Any plan proposed in any Great Lake state or province that involves major consumptive use or diversion must give prior notice to, and seek approval of, all other states and provinces. This principle was reinforced in 2001 in an annex to the charter that commits the parties to develop, by 2004, binding regulations to ensure no net loss to the waters through diversion or consumption or through adverse impacts on water quality. The charter established a common information base where scientific data from all five lakes could be stored and accessed. Like the major cross-national agreement, however, its implementation is dependent upon the voluntary cooperation of the parties involved.

Areas of Concern

The 1978 GLWQA was the first formal recognition of an ecosystem approach to the management of the Great Lakes, and it was advocated by the Research Advisory Board (RAB) in its 1977 annual report to the IJC. The RAB spoke of ecosystems as including human beings, thus providing an explicit metaphysical view of the inseparability of humans and nature. The primary goal in restoring an ecosystem is to sustain those beneficial uses that were impaired by pollution. A list of these use impairments was applied to different sites on the Great Lakes nearshores and connecting channels, the result being the identification of the 43 (later 44) Areas of Concern. See Chapter 1 for the list of impaired uses (see also Figure 1.2 on page 7).

The list of use impairments was drawn up by a binational ecosystems work group consisting largely of government biologists and engineers. It sought public input on the objectives of restoration indicators, but it appears to have been "intellectually captured" by the data already available for scrutiny. Consequently, its purely inductive approach omitted any concern with the sustainability of, for example, commercial shipping and recreational boating and its impacts. These omissions later proved to be a practical hazard with regard to the function and implementation of remedial action plans.

The IJC's RAP program was formally recommended to the two national governments in 1985. They recognized that all government environmental management programs, as well as multiple agencies, jurisdictions, and communities, could be integrated in each AOC (Hartig and Thomas 1988). However, the boundaries of AOCs were not set by particular ecosystem characteristics but, rather, by how use improvement data had been collected. Again, this proved to be difficult to implement across communities that had relatively few social connections with each other (e.g., across national boundaries in the St. Lawrence and Niagara Rivers).

In the 1985 Water Quality Board Report to the IJC, the board stated that it was not clear how the IJC would go about tracking the progress of each AOC. In order to overcome this difficulty, the IJC adopted a system of categories that logically follows the progress of the RAP until its delisting. These categories are as follows:

(1) Causative factors are unknown and there is no investigative program under way.
(2) Causative factors are unknown and an investigative program is under way.
(3) Causative factors are known, but remedial action plan not developed.
(4) Causative factors are known and remedial action plan developed, but remedial measures not fully implemented.
(5) Causative factors are known, remedial action plan developed, and all remedial measures identified in the plan have been implemented.
(6) Confirmation that uses have been restored and deletion as an AOC (in the next Board report) (GLWQB 1985, 45).

The IJC reviews and comments on the plans after Items (4), (5), and (6) above, called Stages 1, 2, and 3. The first stage involves problem definition; the second stage involves the planning and implementation of remedial and regulatory measures; and the third stage involves the monitoring and restoration of impaired beneficial uses. In 1989, the IJC's Water Quality Board reviewed the first eight RAPs submitted. Over half were flagged as inadequate at Stage 1 (problem definition). Some AOCs are still at the Stage 1 level, although most are now at the Stage 2 level. Only one AOC has been delisted – Collingwood Harbour. Collingwood Harbour had relatively minimal pollution problems, and they were solved mainly by updating the sewage treatment plant, curbing agricultural runoff, and dredging contaminated sediments (GLWQB 1991b).

At the start of the RAP program, the three-stage approach (problem definition, planning and implementation, restoration and monitoring) appeared to be straightforward. However, in practice, the three-step RAP is often difficult to implement. Stage 2 of the RAP "is often complex and very lengthy due to the diversity and severity of some use impairments, and the inability to resolve all the problems in the same timeframe" (Hartig and Law 1994, 6). The IJC must rely upon the various American and Canadian governments (federal and state/provincial) to comply with their requirements, including implementation. This makes the geographical scale of the AOCs subject to large areas of political jurisdictions that have incentives to respond to concerns other than those pertaining to ecosystem restoration.

As a result, different jurisdictions have, over time, responded differently to the RAP process. Michigan, for example, has withdrawn from the three-stage IJC process but will issue biennial documents that report on restoration projects that would have been included in Stages 1 through 3 of any of their AOCs (Hartig and Law 1994, 7). The State Department of Environmental Quality, which used to act as lead agency for coordinating the implementation of RAPs in each AOC, is now devolving this role onto the various public advisory committees in each AOC (IJC 1997, Draft, 15-19). (The roles of each committee are discussed in Chapter 5.) Ontario, as a second example (and the sole Canadian province involved with RAPs), has acted in a similar way, ostensibly because of budget constraints. Initially, it moved to endorse part of the Stage 2 documents (those parts recommending actions) but reserved separate judgment on the parts dealing with commitments and endorsements (Hartig and Law 1994, 7). In 1997, Ontario withdrew from its coordinating role in all of its AOCs and connecting channels, leaving Environment Canada as the sole lead agency in the three (now four) AOCs where it assumed the initiative. Both of these examples indicate the institutional weakness of the IJC, which remains dependent upon the cooperation of national and subnational governments for the implementation of its recommendations. Within the Canadian context, as we shall see, there is an equal readiness on the part of one level of government to unilaterally renege on formal agreements. In the 1994 Canada-Ontario Agreement, for example, the federal and Ontario governments agreed to restore 60 percent of all impaired beneficial uses in the 17 domestic AOCs, leading to a delisting (Stage 3) of nine AOCs by the year 2000. The 2002 agreement is now expected to abandon this particular commitment to what was thought to be a model federal-provincial agreement – one that had been successfully negotiated and renegotiated since 1971 (Inscho and Durfee 1995).

Recent Basin-wide Arrangements
The IJC has, in the 1990s, been at the forefront of international efforts to deal with two major concerns on the Great Lakes: (1) the release of toxic substances into the ecosystem and (2) difficulties specific to a particular lake, such as those specific to Lake Erie in the 1960s and 1970s.

In 1990, the IJC began to develop a strategy for the virtual elimination of persistent toxic substances. This followed a 1986 agreement at the state-provincial level known as the Great Lakes Toxic Substances Control Agreement. This agreement was designed to produce more uniform programs and regulations among the subnational jurisdictions on the lakes. It was also a response to its own Water Quality Board's 1985 selection of 11 critical pollutants. These initiatives helped to formulate American and Canadian efforts (undertaken through their respective domestic enabling legislation) to inventory, prevent the release of, and regulate the dispersion of persistent toxins.

These initiatives were also the backdrop for the 1997 Great Lakes Binational Toxics Strategy signed by the United States and Canada. This strategy poses separate administrative actions to virtually eliminate persistent toxic substances by managing them through a lifecycle approach to pollution that would enable pollutants to "achieve naturally occurring levels" (Environment Canada/US EPA 1997, 1). Such a strategy also appears designed to virtually eliminate a history of pollution and a philosophy of empiricism! "Naturally occurring levels" of toxins is, at best, an ambiguous concept. Suffice it to say that the IJC's virtual elimination strategy has made it a focal point of interest group lobbying.

The concern with toxic chemical pollution in the lakes has led national and state/provincial governments to devise lakewide management plans (LaMPs). These are designed to

(1) integrate Federal, State, Provincial and local programs to reduce loadings of toxic substances from both point and non-point sources; (2) assess whether these programs will ensure attainment of water quality standards and designated beneficial uses; and (3) recommend any media specific program actions or enhancements to reduce toxic loading in waters currently not attaining water quality standards and/or designated beneficial uses (EPA 1993, 1).

The LaMPs are supposed to emphasize the reduction of point and non-point sources of critical pollutants. This strategy uses the three-stage process that the IJC adopted for RAPs, and, as a framework, it also uses the IJC list of impaired beneficial uses. Currently only Stage 1 plans (problem definition) have been, or are in the process of being, compiled.

It is fair to state that the IJC's newer programs (those instituted in the 1990s) have not captured the full support of governmental partners with regard to maintaining environmental management on a basin- or lakewide basis.

Supplementary Institutions

A number of supplementary international bodies help to foster integrated environmental management on the Great Lakes, usually on a functional basis. Table 3.1 lists these supplementary institutions, some of which were mentioned in previous chapters.

Table 3.1

Binational governance arrangements outside the Great Lakes Agreement involving institutions independent of the IJC

Institution	Purpose	Members	Activities/History	Staff/Finances
Great Lakes Fishery Commission	Coordinate maintenance of fisheries	4 from each side, named by Privy Council and President	Control sea lamprey; coordinate and advise on other fishery matters	Lamprey costs splits 69%/31% US/Canada; other costs evenly
Council of Great Lakes Governors	Provide a forum on mutual interests	Governors, with premiers as associate members	Developed Great Lakes Charter and seek to promote economic development in region	$20,000 annual dues, plus foundation and private support for special projects
Great Lakes/St. Lawrence Maritime Forum	Promote trade and commerce	Includes government and nongovernment organizations	Promote use of Seaway but has no formal agenda	Funds raised ad hoc for projects
International Association of Great Lakes Ports	Promote Great Lakes shipping	4 US, 5 Canadian port authorities	Lobby on impediments to use of Seaway	Annual dues of $500
Niagara River Toxics Committee	Investigate toxic chemical problems	2 each EPA, NY, Ontario, and Environment Canada	Formed by agencies to recommend actions on Niagara toxics	Staffed and financed by initiating agencies

▼ *Table 3.1*

Institution	Purpose	Members	Activities/History	Staff/Finances
Upper Great Lakes Connecting Channels Study Committee	Assess toxics in rivers and Lake St. Clair	Fisheries and Environment agencies, with IJC observer	Formed in 1984, with study to be completed in 1988	Staffed and financed by initiating agencies
Coordinating Committee on Hydraulic and Hydrologic Data	Coordinate methodology for data collection	Environment Canada, Fisheries and Ocean Corps, and National Oceanographic and Atmospheric Administration	Formed in 1953 to assure compatibility of data	Staffed and financed by initiating agencies
Michigan-Ontario Transboundary Air Pollution Committee	Develop cooperative program for air pollution	Wayne County, Michigan Department of Natural Resources, and 2 from Ontario Ministry of Environment	Initiated by governors and premiers; worked closely with IJC air board to 1983	Staffed and financed by participating agencies
Memorandum of Intent on Transboundary Air Pollution	Develop basis for negotiating agreement especially on acid rain	Government scientists organized in 4 technical working groups	Committee work stalled, with negotiations now by formal diplomatic procedures	Expenses covered by governments through participating agencies

Organization	Function	Membership	Status	Financing
Migratory Birds Convention	Control killing of migratory birds	No formal body for implementation	Signed 1916	
International Migratory Birds Committee	Foster cooperation under 1916 convention	Resource ministers and cabinet secretaries	Established 1960s; has not met since 1970s	
Canada-U.S. Programme Review Committee	Advise governments on protection of migratory birds	3 each from federal governments	Developing North American Waterfowl Management Plan	Research and participation financed by agencies
Mississippi Flyway Council	Recommend hunt regulations	1 from each state and province	Recommend regulations to federal governments	Staffed and financed by participating agencies
St. Lawrence Seaway Authority and Development Corp.	Coordinate construction operation of seaway	Administrators appointed by federal governments.	Determine policies jointly for separate implementation	95% financed by tolls; balance by federal transportation agencies
Seaway International Bridge Corp.	Operate bridge at Cornwall	8 members, mostly from Canada	Maintain bridges and collects tolls	95% by tolls; balance by Seaway agencies
International Boards of Control (4)	Assist IJC decision on levels and flows	Equal members from each side named by IJC commissioners	Develop and implement regulation plans since 1909	Staffed by agencies; report publications financed by IJC

▼ *Table 3.1*

Institution	Purpose	Members	Activities/History	Staff/Finances
International Great Lakes Levels Advisory Board	Advise IJC on levels and public information	16 members, 8 per side, with half the members from public	Carry out studies; reports twice a year	Financed by agencies and IJC
International Great Lakes Technical Info Network Board	Study adequacy of levels and flows measurements	Environment Canada, Fisheries and Oceans Corps, and NOAA	Reported to IJC 1984 on user needs and adequacy of data	Financed by agencies involved in study and data collection
International Air Pollution Board	Advise governments on air quality	EPA, 1 NY, and 3 Environment Canada	Report twice yearly on transboundary pollution	
Joint Response Team for Great Lakes	Cleanup of oil/ hazardous materials spills	Canada and US Coast Guards and other agencies	Maintain Joint Contingency Plan, invoked 9 times since 1971	Staffed by agencies; cleanup costs where spill occurs

Domestic Arrangements

United States

The shared nature of the Great Lakes basin ensures that it relies largely upon Canadian and American domestic environmental legislation to limit waste disposal and pollution and to implement ecosystem approaches to pollution control, such as those embodied in RAPs.

US policy, unlike Canadian policy, is largely dominated by the legislative enactments of the federal government. The main institutional vehicle for such centralization is the judicial interpretation of Section 8 of the Constitution – the commerce clause – which enables the federal government to formulate regulations that may be exclusively contained within just one state.

One key piece of this federal legislation is the Clean Water Act, 1987. Previously, this was the Water Pollution Control Act, 1948 (modified in 1956, 1965, 1972, 1981, and 1987), and it originated in the Refuse Act, 1899 (Freeman 1990). The Refuse Act prohibited the dumping of dredge spoils and refuse into navigable waterways. Thus the common law doctrine of navigable servitude was enlarged by statute (Sproule-Jones 1993). The act now specifies national water quality goals and, through the Environmental Protection Agency (EPA), prescribes standards and priorities for effluent discharged into water bodies like those of the Great Lakes basin. These standards are to be monitored and enforced by state agencies. The act distinguishes between conventional pollutants, such as organic matter and suspended solids, and so-called water toxic pollutants.

The control of toxic wastes – a major concern of the 1990s – is implemented under the Toxic Substance Control Act, 1976, which governs the manufacture, use, and disposal of toxic chemicals. Regulations, again implemented by state agencies under the scrutiny of the EPA, exist for pesticides under the Federal Insecticide, Fungicide and Rodenticide Act, 1974. The Resource Conservation and Recovery Act, 1970 (and subsequent amendments), authorized the trading of hazardous wastes from manufacture to disposal; however, the hazardous waste site issue highlighted by Love Canal (see Chapter 2) drew a specific statutory response. This was the so-called Superfund Act, formerly known as the Comprehensive Environmental Response, Compensation and Liability Act, 1980, which earmarked $1.6 billion (in its initial years) to clean up hazardous waste sites. (The act was not renewed in 1995.) The EPA's efforts to recover cleanup costs from operators of waste dumps has proved to be difficult (Kettl 1993).

RAPs are implemented through the Great Lakes Critical Programs Act, 1990, which required that, by 1992, a RAP be developed for each AOC (Gunther-Zimmerman 1994). It also requires that all RAPs be included in each state's water quality management plan (GLWQB 1991a). The American RAP requirements are thus spelled out in law, unlike the Canadian RAP requirements, which are a product of an (executive) Canada-Ontario agreement.

The National Environmental Protection Act, 1969, was, in part, a direct response to the Lake Erie pollution. Under this act, and its various state equivalents, any major action significantly affecting the quality of the environment requires an environmental impact statement. This procedural requirement is enforceable by citizen suit, and the Council on Environmental Quality provides oversight at the federal level (Rogers 1996, 98). However, the specific incentives, both regulatory and financial, for the Lake Erie cleanup came from a predecessor of the Clean Water Act.

A plethora of federal and state legislation supplements these key acts. These include a Safe Drinking Water Act, 1974, amended in 1986, that regulates 83 potential water contaminants. They also include acts that deal with indirect impacts, such as those that occur through non-point source pollution. Rogers lists 35 federal agencies in 10 Cabinet departments, 11 independent agencies, 4 agencies in the Executive Office of the President, a number of federal courts, and 2 bureaus that have some responsibility for water programs and projects on the Great Lakes (Rogers 1996, 99). The EPA, however, remains the central agency around which other federal, state, and local bodies cluster. Thus, compared to its Canadian counterparts, American environmental legislation is legalistic and centralized (see Hoberg 1993; Howlett 1994; Johns 2000).

Canada
Both the Canadian and Ontario governments have major roles to play in the environmental management of the Great Lakes. They can enter into extensive negotiations to minimize duplication or, alternatively, to enlarge their management responsibilities in response to sectoral or public interest group pressures (Sproule-Jones 1993). Their precise roles thus change over time and from one policy field to another. In the Great Lakes basin, where multiple uses and policies abound, both provincial and federal interests have potential co-equal status to restore and enhance ecosystems.

The basis for the co-equal status is the Constitution Act, 1867, within which various provisions in Section 92 grant legislative authority to the provinces. This is especially true of Clause 13, the property and civil rights section. Proprietary jurisdiction over water resources is allocated to the province under Section 109. Federal authority derives mainly from Section 91, the "peace, order and good government" preamble, the fisheries clause (12), the navigation and shipping clause (10), the criminal powers clause (27), and the treaty power under Section 109.

The structure of parliamentary government at both the provincial and federal levels creates incentives for the party in power to keep legislative enactments relatively broad and to spell out policy directions in statutory instruments (regulations) and policy manuals. These, in turn, often emanate in closed-door negotiations between regulators of waste dischargers and the dischargers themselves (industries and municipalities). It also circumscribes judicial review.

RAPs are formulated by teams of provincial and federal officials with the (varying) aid of different groups of stakeholders in different AOCs. They are administrative arrangements formulated under the Canada-Ontario Agreement (originally 1985) and are not considered to be legal contracts. RAPs, like LaMPs, must be implemented under specific pieces of provincial and/or federal legislation.

The core implementing legislation consists of the Province of Ontario's Environmental Protection Act, 1971 (as amended); the federal Fisheries Act, 1989; and the Canadian Environmental Protection Act, 1988. The Ontario EPA is used to impose effluent standards on all stationary point sources discharging into the basin. The standards are so-called scientific objectives, and the relevant regulator (the Ministry of the Environment) can bargain both the degree and timing of compliance with these standards. Since 1986, the province has gradually been introducing total loading standards per type of waste rather than the dilution standards embodied in effluent streams. At present, total loading standards are implemented only for the oil and gas and the iron and steel sectors, and there is some uncertainty as to whether they will be extended to other sectors.

The Fisheries Act, federal legislation that was first enacted in 1868, can be implemented by either federal or provincial authorities under a century-old interdelegation of powers, again accomplished by administrative agreement. Its value lies in protecting the assimilative capabilities of water, as set out in its famous Section 36:

> No person shall deposit or permit the deposit of any deleterious substance of any type in water frequented by fish or in any place under any conditions where such deleterious substance or any other deleterious substance that results from the deposit of such deleterious substance may enter any said water (RSC, F-14, S 36(3); 1989).

The Fisheries Act is both a bargaining tool for regulators and an occasional means of justifying the prosecution of waste dischargers.

The Canadian Environmental Protection Act consolidated a number of previous federal statutes and extended federal authority to regulate the import, export, transport, distribution, and use of designated toxic substances. It can thus be used to implement any virtual elimination strategy formulated by the IJC or by the joint agreement of Canada and the United States (see above). Currently, both Environment Canada (and the US Environmental Protection Agency) monitor a voluntary reduction of different persistent toxic substances on the part of both users and dischargers.

As in the United States, so in Canada there is a plethora of other statutes that cluster around these three major statutes and that describe the institutional bases for restoring the Great Lakes. These include the Ontario Water Resources Act, which vests all riparian rights to the uses of water in the provincial Crown (Campbell, Scott, and Pearse 1974), thus giving the government the legal authority to regulate water levels and pollution (primarily from non-point sources). Ontario's Environmental Assessment Act adds pre-project approval mechanisms for major private-sector and all public-sector undertakings (Gibson and Savan 1986). A parallel federal process exists under the Canadian Environmental Assessment Act, 1995. The Conservation Authorities Act, 1970 (Ontario), enables these provincial Crown corporations to regulate stream flows and erosion controls (with municipalities) on a watershed basis (Mitchell and Shrubsole 1992). Finally, Canada, like the United States, has a national Navigable Waters Protection Act that prohibits foreshore uses that could inhibit shipping, even (in the US case) on non-navigable wetlands (Rogers 1996, 95).

On the Canadian side of the Great Lakes basin, therefore, restoration of the Great Lakes is just as likely to rely on provincial regulations and management of resources as it is on federal regulations (Lucas 1990; Cairns 1992). The precise configuration of statutes and regulations that apply to pollution controls has to be unravelled both on a discharge-by-discharge basis (for point sources) and on a site-by-site basis. This means that Canadian government departments and agencies have more discretion than do their US counterparts. However, because of the constitutional centralization of power in US federal bodies, the US may be able to act with greater legal consistency across basin sites than can Canada. Current Canadian policy is to develop comprehensive framework agreements between the Canadian government and each province in order to reduce the transaction costs of governance and, perhaps, to provide consistency.

Conclusions

For over a century, the communities surrounding the Great Lakes have attempted to ameliorate and control some of the negative effects of rapid urbanization and industrialization. In the last 30 years, governments have made a concerted effort to control waste discharges, regulate resource uses, and restore degraded ecosystems. During this time, much of the intellectual leadership has been provided by the IJC.

Recent efforts at management and restoration have focused on RAPs for the severely impaired AOCs. LaMPs and new toxic control strategies have supplemented the RAP processes. They, in turn, have built upon systems of pollution control that require the dilution of most liquid wastes before disposal in the lakes. Non-point source controls on wastes, in situ treatments (or dredging and treatments) of contaminated dumps and sediments, and lifecycle management of hazardous materials remain current "to-do" items on government agendas. There is a remarkable level of convergence in the Canadian and American policy strategies for the Great Lakes.

Due to their different constitutions and cultures, the precise institutional arrangements and policy styles favoured by Canada and the United states do vary. The United States has a more centralized and judicialized process of governance than does Canada, while the latter has a more bureaucracy-driven and closed governance process than does the former. Both countries, of course, share comparable common-law legal inheritances.

We have described the basic framework of legal rules devised to manage and restore the Great Lakes ecosystem, and it is now time to move from the formal legal framework of rules to the actual application of rules in the real world. Informal rules, conventions, operating procedures, and general agreements between users and user organizations supplement formal statutory rules to provide a real nest for restoration purposes. In Chapter 4, we explain their character and the incentives they provide to individuals, organizations, and groups. We draw upon common property resource theory in order to extend the analysis to the state of the Great Lakes and the institutional experiments known as RAPs, and we develop a framework within which to systematically, and cross-culturally, compare one RAP with another. In Chapter 5, we look at RAPs in light of the framework developed in Chapter 4. In Chapter 6, we combine theory and practice.

4
Common Pools and Multiple Uses

Common pools are assets that are subject to the subtraction principle. This simply means that, once some of the asset is taken, then there is less left to take. Common pools are like refrigerators: you may fill them with food, only to find that others in your household have taken much of it and little is left for you. Unless something is done about this, the refrigerator may soon be empty.

Common pools are ubiquitous. They can be given by nature (e.g., a well, an oil pool, or a living but hunted animal species); they can be socially constructed (e.g., a government budget or a well-stocked refrigerator); or they can be a combination of both (e.g., a harbour to dump human and industrial wastes).

Gradually, scholars are beginning to understand and to solve (at least intellectually) the ubiquity and problems of common pools. In the real world too, communities working alone or with government help are beginning to understand and solve the basic logic of common pools. What is the basic logic? Do common pools amount to anything more than a replenishment issue? Can we solve our refrigerator problem by simply restocking it regularly?

One obvious problem is that, aside from having to go to and from the store (a kind of transaction cost), it gets to be quite costly to keep on refilling the refrigerator. One may not have the resources to keep up the replenishment at the rate at which the stock is being reduced. We may also feel that it is unjust that others are allowed to reduce the stock that only we are prepared to replenish. In some cases, this injustice can mean that we will allow the stock to be exhausted rather than constantly allow ourselves to be "suckered" by others.

In any case, given our ignorance of the technical nature of the stock and/or our inability to restock properly, many common pools cannot be replenished. The natural world provides the best examples of these common-pool problems. We can restock some fisheries, for example, with hatcheries and laboratory fish. But captivity-bred fish do not exhibit behaviour identical to that of fish reared in the wild, and their resilience in natural waters, let alone their taste to fishers, can be quite different from that of wild fish. Nature by itself may restock well, but this process may take so long that we treat the resource as non-renewable (think of an oil pool, for example).

So with these limitations, we may decide to devise rules to limit the use of, or withdrawals from, the common pool in question. Rules can, perhaps, be fashioned so that the rates of withdrawal are approximately equal to the rates of natural replenishment (and/or artificial replenishment). If one can limit the number of times a refrigerator door can be opened, then perhaps, with some judicious shopping, the stock of food will stay about the same. One may also need a cluster of social norms to make this work. Frequent comments about greedy people, freeloaders, or the trials of modern shopping could be a socially productive use of human guilt. There is a wealth of evidence that small communities can fashion enforceable social rules that enable them to sustain many common pools over time (McCay and Acheson 1987; Pinkerton 1989; E. Ostrom 1990; McKean 1992; Bromley et al. 1992; also, E. Ostrom, Gardner, and Walker 1994 provide major reviews). However, it is not easy to construct rules for all common pools in all social and natural environments. One reason for this involves the technical characteristics of different pools.

Dimensions of Pools

We seem to have studied common-property resources in physical-biological situations (such as in rivers and lakes) more than we have studied them in non-social situations (such as government agencies and committees). My examples tend to be drawn from the natural resources and environmental area.

It is well recognized in the literature that a key dimension of common pools involves exclusion. It can be difficult to exclude some persons from catching or using a resource. Migratory fish, for example, have a way of avoiding international boundaries and the rules devised to limit fishing effort and to sustain the resource. Consequently, fishers from neighbouring countries may enjoy what you thought were "your fish." It is the technical issue of fugitivity that makes exclusion difficult. Similarly, a common pool used for dumping liquid wastes may successfully assimilate oxygen-demanding wastes through the regular mixing of its waters with the atmosphere. However, the rates of assimilation may be too slow to deal with extra loadings that might come from unexcluded sources (e.g., upstream erosion on a river). Again, there are technical characteristics of the good or resource that may make exclusion difficult. These technical characteristics may not only make exclusion difficult, but they may also make consumption or withdrawals from the pool unpredictable and/or replenishment rates inaccurate.

Rules for excluding users of a common pool may be insufficient or inappropriate. For example, rules that tightly regulate the kind and quantity of wastes pumped into a harbour may be insufficient if no rules exist to regulate such non-point sources of pollution as storm-water runoff. Rules designed to control human sewage wastes may be, and frequently are, inappropriate for dealing with persistent organic chemicals discharged into the environment. A lot of the rules that communities and governments fashioned in order to control the environment were (and often still are) modelled on the Benthamite public health movement of mid-nineteenth-century England. They involved designing sewers, disinfecting wastes, and discharging the resultant effluent far from population centres. Rules that were once necessary and sufficient became, with the development of industrial capitalist societies, necessary but insufficient. Thus technology is also a defining dimension of common pools.

A third key dimension of common pools such as the Great Lakes involves interdependence. Living systems are interdependent in differing scales (and times), depending upon their functional relationships to each other. This is the primary characteristic of ecosystems (de Groot 1986). So, as a crude example, there is a functional interdependence between the regulation of nutrients in wetlands, the productivity of wetlands for aquatic animals, and the carrying capacity of the natural environment for human food sources. Stresses on any one of these relationships, such as would be caused by massive waste disposal, can threaten the resilience and long-run sustainability of the ecosystem as a whole. It is the scale and variety of these interdependencies that give an environmental common pool an extra level of complexity and uncertainty.

Dynamics

Common pools change and adapt over time as a result of changes in any one of their defining dimensions: exclusionary rules, technology, and ecosystem interdependencies (in the natural resources cases). The economic literature on common pools emphasizes and differentiates between the consequences of changes in the stock (or size or population) of the pool and the consequences of changes in the intensity of its use (within the parameters of zero technological change and zero ecosystem connections). A frequent result of this type of situation is the so-called "tragedy of the commons," whereby common-pool users increase their intensities of use beyond the sustainable range, with the result that the stock of the pool is reduced (if not eliminated). This particular scenario, which emphasizes the incentives for users of a resource to become opportunists, is the subject of extensive research. It has also been modelled as the so-called "Prisoner's Dilemma Game" in the formal branch of economics known as game theory. However, it is but one possibility in a range of possibilities.

Let us turn again to the example of the household refrigerator. The stock is its contents; the flow is the set of items put in and taken out by the household members at any one time. The technological conditions are set: one refrigerator and one source of edible food. The exclusionary rules are also set: go to the refrigerator whenever you want and remove however much you wish. There are no real non-trivial ecosystem interdependencies in this case. The incentive for you as a household member? Withdraw items for consumption (or for storage and later consumption in case your roommates/partners/family members get greedy and eat into the stock). Other household members who think the same as you will similarly reduce the stock. The collective interest of rationing the resources until the stock gets replenished is ignored. What a tragedy!

Economics often uses static equilibrium models to understand common pools and their limits; this directs attention away from their dynamic character and their changing parameters. Early work pertaining to understanding ecosystems had the same emphasis; namely, a concern that environmental factors set limits on plant and animal populations. In both social and environmental sciences in recent years, there has been a renewal of concern with dynamic changes in pools. Evidence exists that the tragedy of the commons is a relatively rare situation and that, over time, communities tend to develop exclusionary rules to help reduce dynamic fluctuations in common-pool populations (E. Ostrom 1990; E. Ostrom, Gardner, and Walker 1994). Evidence also exists that management of fluctuations of populations is prone to failure because of an inability or unwillingness to manage more than a selected set of variables (Francis 1994; Lee 1993; Gunderson, Holling, and Light 1995). The result can be major surprises (Holling 1986). Fish populations, for example, kept high by hatchery stocks can suddenly plunge as the less-adaptable hatchery fish succumb to disease. Long-run dynamic adaptations were not understood and thus, in part, were ignored in initial decision making regarding the sustainability of the resource.

A major reorientation towards the dynamics of common pools is necessary. Some work and investigations of property rights in the natural resources field suggest that an effort to understand the dynamics of rights (and, more formally, rules) may help to reveal more reliable methods for sustaining common pools. We will now review and then extend current understandings about property rights and their adaptations over time. We discover that our knowledge is rooted in our understandings of individuals as private property owners, proprietors, managers, and users. We thus explore extending the logic associated with these situations into situations where the stakeholders may be corporate and collective organizations, and where, as a result, different methods of adaptation are developed. Our focus remains on the conditions (or rules) through which changes are made. These conditions are termed collective-choice rules, and they determine, in part, how property rights, and what we have been calling "operational rules," change and adapt. It is within these collective-choice arrangements that rules about common pools are formulated and implemented by individual and collective interests. If we are to understand how ecosystems evolve, develop, and change, then we must analyze how collective-choice mechanisms operate and adapt. In order to do this we present a basic framework of rules analysis, including collective-choice mechanisms, and extend it to deal with major classes of factors. Chapter 5 develops indicators to test this framework within the context of AOCs on the Great Lakes.

Property Rights
Property rights are "legally sanctioned" rules that affect the uses of resources and "the corresponding assignment of costs and benefits" (Libecap 1986, 229). The resources in question include physical possessions, like land, or dwellings, or equipment. They can include knowledge as a resource or, in unusual cases, other human beings (e.g., conscripts in a standing army). Property rights are, in fact, "bundles" of rights that can include the right to exclude others, which we saw as a key dimension of common pools. Other property rights include access, withdrawal, management, and alienation (transferability) (Schlager and Ostrom 1992). Different configurations of these rights are found for different resources. If we use our refrigerator example again, then we see

that different households may place different limitations upon access. These would be tailored to the different demands for, and supply of, food.

Of particular concern for this study is the structure of control and power over these resources. Who can ration the food? When, where, and how? Existing theory identifies three classes of persons who can exercise property rights. There is the individual person, like the "head" of the household, who can set the rules for access, withdrawal, management, exclusion, and transferability regarding the food in the refrigerator. Besides individual property rights, we also find that some rights are held collectively (e.g., fishers may collectively control various uses of a fishery). Finally, we can find resources that are not held by anyone. Many fugitive natural resources are of this character and form the best cases for understanding the tragedy of the commons. No one can exclude others in such situations.

This threefold division may be useful in describing many situations, but it brings little understanding to real-world circumstances. It is like saying there are only three ways to husband a resource; namely, to put one person in charge, to put everyone in charge, or to put no one in charge. Most households would fail if presented with only these three choices regarding how to monitor their refrigerators.

What we can do is disentangle the concept of property rights in order to distinguish between different kinds of "stakeholders." Each of these kinds of stakeholders possesses some degree of control over a bundle of property rights, but each may be subject to different degrees of transaction costs; that is, they vary in their abilities to adapt to new circumstances.

At one end of the dimension of stakeholder types one can speak of individual property stakeholding, where, for example, a person can own, lease, manage, or use a resource for his or her purposes. Many formal economic models of resources rest on this kind of arrangement. It is also the model for many legal analyses of property rights, and it is the intellectual paradigm for the practical formulation of the common law on property (LaForest 1969; Berman 1983). Individual property ownership clearly specifies who is the stakeholder. One can assume that the transaction costs involved in changing the modus operandi of the individual in question tend to be close to zero (Hamlet was an exception). The merit of this model may lie in its adaptive capabilities as well as in its clear specification of ownership.

The change in the bundle of rights that an individual property owner may possess, however, is partly a function of wider social processes.

Vernon Smith, in an original effort to discover the origins of and changes in property rights, suggests that certain social conditions that were conducive to property rights occurred in the late stages of Neanderthal and the early stages of Cro-Magnon Hominid periods (roughly from 90,000 to 100,000 years ago) (Smith 1993, 169). The first condition involved the trading of valuables, which requires some property rights and contracts to be in place so that traders can trust the exchange process. The second condition involved adopting a more sedentary lifestyle, which was made possible by the secure accumulation of possessions and knowledge of their acquisition. From these two conditions, there developed a form of customary law that protected against damages to property and persons (Benson 1990, 13-14). At first, this law was enforced by kinship networks, then by dispute settlement forums. Settlements could be voluntary (e.g., those agreed upon before peers in the merchant courts) or authoritarian (e.g., those developed in Anglo-Saxon courts for breaking "the King's peace"). Customary law provided for a voluntary adaptation of property rights (and duties), while authoritarian law permitted a more radical form of change. The latter is the type of law to which most governments aspire.

Adaptation and change can occur more readily when private property owners possess the legal right to transfer their resources in the economic marketplace. These legal rights can be attenuated. The common law may or may not lend itself to the continual adjustment of these legal property rights. Consider the following extended example.

Adaptations in Property Rights: An Extended Example
In common law jurisdictions, flowing or running waters cannot be owned. The use of water may constitute a bundle of legal rights rather than the possession of water per se. This bundle of legal rights to use water is, in common law, a derivative of the right of ownership of land. A riparian is one whose land is washed by water and who has the right to the natural flow and quality of that water, subject to the same rights as his riparian neighbour: "No riparian proprietor has any property in the water itself, except in the particular portion which he may choose to abstract from the stream to take into his possession, and that during the time of his possession only" (Lambden and De Rijcke, S35, 148-54). The riparian is thus enjoined to make reasonable use of (surface) waters, subject to the same reasonable use by his neighbours. However, if the water is diminished by the reasonable use of upstream riparians, then the downstream riparians have no legal recourse. Reasonable use is determined through the judicial process. In common law, it does not

include the right to pollute or to obstruct and divert the waters, nor does it include the right to remove unreasonably large quantities (unless so granted by other riparians).

Riparian rights are transferable in that a non-riparian may negotiate access to the water, subject to the uses of other riparians. However, "in order to obtain secure rights to flows, a non-riparian is likely to have to acquire rights from all downstream and all upstream riparians in order to secure his supply" (Campbell, Scott, and Pearse 1974, 481). A riparian may also negotiate rights to an extraordinary access to water. Again, the agreement of all riparians is necessary since any one of them can claim compensation for a reduced flow resulting from extraordinary use. The riparian need not exercise her/his right in order to preserve it and, in fact, is entitled to the natural beauty of the water itself.

Thus, in common law, the use of surface waters for domestic, farming, or other purposes is contingent upon riparian ownership of adjacent lands. These uses may undergo continual change as a result of judicial determinations of reasonable use or through bargaining between riparians, through market negotiations between non-riparians and riparians, or through purchase of riparian lands on the open market. The property rights themselves are made adaptable through continual adjustments, as represented by the concept of "reasonable use."

The common law on groundwater, peculiarly, is not subject to "reasonable use" but is subject to the rule of capture by any landowner with physical access to the acquifer or groundwater basin. While pollution of groundwater is a private nuisance and, hence, if unreasonable, then justiciable, "this percolating water below the surface of the earth is therefore a common reservoir or source in which nobody has, as far as he can, the right of appropriating the whole" (Dubin 1974, 186). Thus groundwater property rights are attached to the land and are not adaptable, although the bundle of rights as a whole may be purchased on the marketplace.

Within this context, governments have often intervened to initiate a process of adaptability in property rights, both for surface use rights and for groundwater rights. Generally, governments have replaced or modified the reasonable use doctrine for surface waters and the rule of capture for groundwater with some form of permit or licence to take water in approved quantities for approved purposes. Thus, in the Province of Ontario, the Ontario Water Resources Act, 1970 (revised 1985), establishes that withdrawals of more than 50,000 litres/day require a permit from the Ministry of the Environment. Such withdrawals are considered to be extraordinary and do not attenuate the riparian rights of reasonable use. However, the director of water resources retains the legal power to limit withdrawals smaller than 50,000 litres/day if they might constitute a nuisance for other landowners. These discretionary powers are authorized under some draconian legislative clauses. Section 11, for instance, states:

> The Minister, for and on behalf of the Crown, may for the purposes of this Act, acquire by purchase, lease or otherwise or, without the consent of the owner, enter upon, take possession of, expropriate and use land and may use the waters of any lake, river, pond, spring or stream as may be considered necessary for his purposes and, upon such terms as he considers proper, may sell, lease or dispose of any land that in his opinion is not necessary for his purposes.

The adaptability of property rights to newer socio-economic and environmental conditions can, at one level, depend on the reasonableness of current usage adopted by the owner or proprietor. Other interests must be taken into account, at least in the case of surface waters. In the case of groundwater, the reasonableness test is limited. However, the common law of water resources provides only some operational rules for day-to-day decision making. Governments may implement statutory legislation that can reallocate decisions or, more fundamentally, re-order the bundles of property rights in water uses. Governmental powers are powers of collective choice; they are rules about rules. To put it more precisely, they are collective rules about operational rules. The characteristic of "rules about rules" will be dealt with later.

Other Stakeholders

Most of the theoretical work and model construction developed in order to understand property rights deals with the private property case, where the owner possesses a full bundle of rights, including transferability. There are, of course, other stakeholders that may possess some property rights, including public enterprises that are major resource owners in both Canada and the United States (Feldman and Goldberg 1987).

Beyond the private property case is a range of organizational (rather than individual) enterprises engaged in decision making about resources uses. The public enterprise, a corporation owned by a government but at arm's length from day-to-day interactions, is one key variant. Another is the private corporation (legally a public corporation, as its ownership is not privately retained but made available through stock and bond acquisitions) that is often engaged in resources extraction. A third is the private non-profit corporation that may acquire some property rights out of its interest in resource uses. Ducks Unlimited and the Sierra Club foster major acquisitions of property rights to land and water out of their own concerns. All three types of corporations are ostensibly formed to facilitate speedy group decision making that is designed to mimic individual decision making. However, all three types experience varying degrees of transaction costs in formulating and implementing actions, as is well documented in organizational theory and behaviour. For our purposes, we need simply note that they are organizational equivalents to the private property stakeholder in that they may exercise one or more of the bundle of rights enjoyed by the individual private person but that they incur higher transaction costs as a consequence of their own "internal collective-choice" arrangements.

Like the individual private property case, all three kinds of corporate stakeholders are subject to societal collective-choice processes with regard to changes in their institutional arrangements. Corporate law is, typically, a mix of statute and common law where the statutory component, at least, provides a framework for specifying the rules that must be followed for the making of the corporation's internal "bylaws." Thus, to some degree adaptation can take place internally; however, the overall framework is subject to statutory legislation and implementation. As legal persons, these organizational forms are subject to the specific tests of resource usage (like, for example, the "reasonableness" test of riparian rights).

Beyond the corporate cases of property rights holders is a third type of stakeholder. This is the non-corporate organization that may have formal institutional arrangements. Major examples are the line departments of government at all levels of formation, from the local through to the national. Due to the complex legislative-bureaucratic arrangements needed to change organizational form and direction, these departments may have fewer opportunities to adapt than does the corporate form. Yet line departments, such as the US government's Department of the Interior and/or the Canadian provincial governments' natural resources departments, may operate on behalf of their "sovereign" governments (or Crowns) with regard to particular (and occasionally specially established) property rights. These stakeholders may all exercise one or more property rights, and all vary in their ease of adaptation, from the individual person (including the squatter) through the corporate person to the non-corporate organization.

Perhaps it is more useful for our purposes to examine the legal positions that stakeholders can hold rather than just their possible bundle

Figure 4.1

Property rights associated with positions

	Owner	Proprietor	Claimant	Authorized user
Access and withdrawal	×	×	×	×
Management	×	×	×	
Exclusion	×	×		
Alienation	×			

Source: Schlager and Ostrom (1992, 252).

of property rights. Legal positions are associated with different sets of property rights. For example, Schlager and Ostrom (1992) use the following typology to assign legal positions to different sets of property rights (Figure 4.1).

Thus an owner can exercise all four major property rights, perhaps individually or perhaps jointly. Similarly, a proprietor can exercise three of these rights but cannot transfer the resource to another legal person. A claimant has further attenuated rights and an authorized user has only the rights of access to and withdrawal from the resource. A squatter is someone who acts as an unauthorized user of land or water.

Empirically, the different kinds of stakeholders can assume any of the four legal positions noted in Figure 4.1. Thus a governmental line department can be found, in some situations, to be an owner with a full bundle of rights or simply an authorized user of a resource. Private property rights can include a full bundle or a lesser bundle. The public/ private distinction between governmental and non-governmental organizations assumes little value in assessing legal positions. On the other hand, as a slowly adaptive stakeholder a governmental line department is less likely to transfer a resource adroitly than is an individual person or a corporate person. This is due to the relative non-adaptability of its exercise of the property rights of alienation. Figure 4.2 arrays the possibilities associated with legal positions and stakeholder adaptability.

Figure 4.2

Legal position and stakeholder adaptability

Legal position	Stakeholder adaptability		
	Individual person	Corporate person	Non-corporate person
Owner			
Proprietor			
Claimant			
Authorized user			

Property rights are rules that have been constructed and used by stakeholders engaged in some form of resource "exploitation." They consist of a bundle of rules and can perhaps be best understood as an adaptive and adapting framework within which decisions about resource usage are taken. To return to our refrigerator example, it is as if a household has drawn up a set of rules about when, how, and under what conditions food may be withdrawn from the refrigerator (and by whom), and when, how, and under what conditions it may be restocked (and by whom). Further, the household may have elaborate rules about adapting the previous operational rules of "exploitation" of the food resource. With a simple head of the household, the rules can be re-adapted relatively easily; with corporate arrangements, which essentially delegate decisions to two or more household members, rules are not easily re-adapted. And non-corporate arrangements, like having ownership "rights" shared equally among a large household, can bring about long and arduous decision making regarding changing the rules for, for example, the use and replenishment of milk cartons or ice-trays. Failure to re-adapt such operational rules could lead to a "tragedy of the milk commons" or to the exploitation of the "sucker" charged with buying more milk. With any given set of situations, their re-adaptation in the light of new circumstances (such as an increase in household numbers) will depend upon the rules set up for handling collective choices.

Collective-Choice Rules

Just as resources and property rights evolve and adapt in dynamic fashion (albeit, in the case of the latter, with lags and leaps), so collective-choice mechanisms may also experience dynamic change. Collective-choice mechanisms are rules about rules: specifically, they are rules about how operational rules are reviewed and changed. Collective-choice mechanisms themselves may be subject to a third level of rules, called constitutional rules. Constitutional rules govern the processes through which collective-choice decisions are taken. They can involve relatively simple frameworks (like the rules governing household decision making) or relatively complex sets of rules (like the national constitutions of Canada and the United States). Rules are thus stacked from the operational level through the collective-choice level to the constitutional level (Figure 4.3).

Some features about rules and rule stacks should be mentioned at this juncture. First, rules do not fully determine outcomes or decisions. Individuals have levels of discretion within which to make choices and to learn from them over time. One may find that, over time, a recurring

Figure 4.3

Rule stacks

Constitutional rules
↓
Collective-choice rules
↓
Operational-choice rules

pattern of choices is both necessary and optimal. In spot markets, for example, with standardized products and low entry costs, producers may discover over time that marginal cost pricing is their most successful choice (as well as being necessary to keep them competitive). On the other hand, with regard to dynamic natural resource situations – like a harbour or river mouth subject to variations in climate, water flows, nutrient loadings, and pollution – owners of a resource such as a fishery have a wider range of choices and a wider range of uncertainties about the outcomes of these choices. Errors are likely to be more frequent. The property rights regime, no matter what its operational construction, will not eliminate these errors. (Although it could, of course, amplify them.)

Second, in practice, many rules at the operational level, collective-choice level, or constitutional level are ignored. They have become simply rules-in-form rather than rules-in-use (Sproule-Jones 1993), or what Walter Bagehot, the Victorian observer of the British Constitution, once called the "dignified" part of the constitution (Bagehot 1964). Some examples may help to clarify the distinction between rules-in-form and rules-in-use. Let's use the refrigerator case. One collective-choice rule regarding stakeholder participation could be that all six members of the household have an equal share in decision making pertaining to replenishment (shopping) of the resource (food). In practice, however, this rule could become a rule-in-form if one person continually takes the initiative (with the compliance and then agreement of others) to decide when, where, what, and how much to purchase. A new rule-in-use has evolved. The constitution of the household permits this evolution because it does not spell out an amending procedure or an enforcement mechanism regarding ongoing collective-choice rules. It may be hard to get all the household members together to redefine an agreement that holds that replenishment decisions can and should be taken by one person rather than by "the collective." These kind of changes recur in complex real-life situations, as we saw in our previous examination of water rights.

Third, considerable time and effort is often expended at the "constitutional level" in formulating precise and practical rules about collective choices. Collective-choice rules would normally comprise rules about the articulation of stakeholder interests (such as selection and representation of stakeholders) and rules for the aggregation of these interests (such as time and frequency of meetings and rules to resolve conflicts). Such rules are critical in establishing the power relationships that could manifest themselves in operational rules and decisions. Allied with these particular collective-choice rules may be rules about the implementation of decisions, by whom and how and with what solutions, as well as outcomes and measures of outcomes. Implementation of decisions in the form of operational rules and policy outcomes is necessary to make all of the governance rules a useful system (pun intended).

Finally, collective-choice rules, like rules at other levels, are contingent upon their peculiar situational conditions. Not only will a resource (or other good) have its own technical characteristics, and perhaps ecosystemic characteristics, but it will also be rooted in particular times and places. These times and places will impose constraints and facilitate opportunities for living systems to evolve and develop. Our trivial refrigerator case is no exception. It will be situated in a household with particular characteristics and demands on the resource, and members of the household will have their own nutritional histories. Moreover, the household will have its geographical location and times of function and change, all of which can have potential effects on refrigerator uses. The pool can change for reasons other than rules or technical discussions or interconnectedness. These situational characteristics always impose limits on the precise applicability of generalizations and hypotheses.

A Basic Framework for Institutional Analysis

The contingencies of time and place may make the precise modelling of real-world common pools difficult. However, they still permit us to

Figure 4.4

Basic framework I for institutional analysis

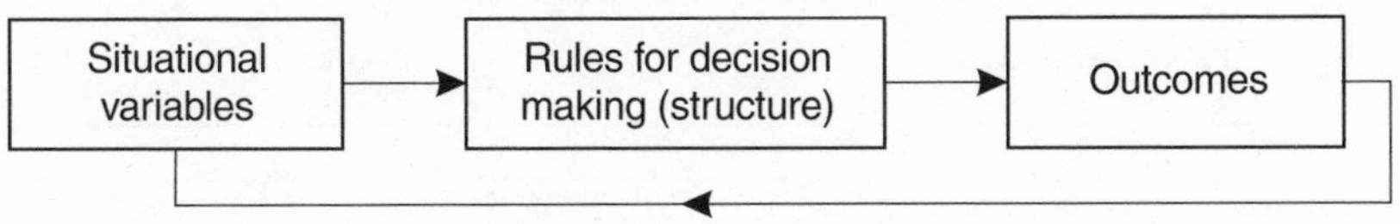

advance a framework linking key factors or variables together rather than a model that specifies the functional relationships between the variables. The basic framework is presented in Figure 4.4.

This simple framework cannot capture the detail of all common-pool characteristics, decisions, and effects of decision making, let alone the sets of rules within which decisions take place. However, it provides us with a skeleton framework that enables us to orient our analyses. Essentially, it hypothesizes that structural variables like property rights can affect ecosystems (or other common pools). However, other situational factors, like the technical characteristics of the resource or the characteristics of human communities that use that resource, can also affect outcomes. These latter relationships are reciprocal too.

We can also sketch the outlines of a slightly more complex framework (Figure 4.5), given our previous discussions of rules. It is expressly applicable to the differing ecosystem and community conditions (the situations) in different AOCs on the Great Lakes. Also, in light of our previous chapter on the US and Canadian constitutional frameworks for environmental governance, it includes the constitutional level of rules. The outcome variables include both implementation process variables (like the use of special agency structures) – which, normally, are not considered outcomes – and actual ecosystem impact variables. The implementation process is distinctly characterized as an outcome variable in this case because many RAPs were formulated with the feasibility of new implementation structures in mind (therefore, by definition, they are not part of the initial collective-choice arrangements). Of course, the entire process gets animated by human motives and energies, and these variables, while endogenous to our concerns, would be included in a formal model rather than in a framework of understanding. Again, we reject the relevance of a formal model due to the variety of situational variables extant throughout the AOCs.

Figure 4.5

Basic framework II for institutional analysis

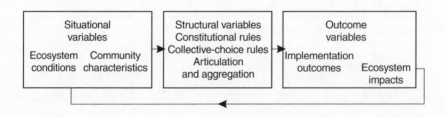

In the next chapter, we will develop measures of these variables or factors in our framework for greater examination in our survey and case studies of Areas of Concern.

Multiple Uses and Multiple Externalities

The basic frameworks for analysis emphasize the centrality of collective-choice processes for understanding dynamic changes in operational rules (like property rights) and ecosystem conditions. They stand in contrast to much orthodox thinking, which emphasizes rules as parameters rather than as evolving tools, and socio-economic factors (such as homeostatic or equilibrating factors). Our frameworks focus upon the very processes involved in the adaptation of rules.

Within the context of previous studies of pollution control, environmental management, and common pools more generally, this approach also contrasts with traditional intellectual and legal foci on simple resource uses and methods to contain negative interdependencies (externalities). These studies embrace models that deal with single, not multiple, uses and with externalities consisting of one or a few other uses of a resource. The most seminal of these studies is the Coase Theorem (Coase 1960).

We have developed no names in the literature to characterize the multiple use and multiple externality problems that are the central issues in many real-world situations (such as AOCs on the Great Lakes). We might characterize our basic frameworks as the "no name" frameworks for addressing the problem of the institutional bases of multiple uses and ecosystem interdependencies. As a matter of speculation, it might be a condition of human society and development that it cannot satisfactorily address real-world concerns until it can find a name to fit them. If correct, then this might imply that metaphysics is a branch of linguistic anthropology, and that is something that raises issues and conjectures well beyond the scope of this study (V. Ostrom 1997).

Conclusion

The governance of common pools involves institutional arrangements, or rules. In natural resources situations, societies have developed customary rules and common law rules – encapsulated in the term "property rights" – to ensure a fair and sustainable way to access, use, and withdraw from the resource (i.e., to exploit it). All rules will need to adapt, however, to meet new situations – physical, biological, and social. So property rights and other so-called operational rules of conduct are subject to wider societal processes of adaptation. These wider societal processes involve collective-choice rules within which stakeholders, possessing varying bundles of property rights, will articulate and aggregate their interests. Decisions will emanate in the forms of revised operational rules and other outcomes. RAPs for the 43 AOCs on the Great Lakes are, in essence, experiments in collective choice.

RAPs are drawn up by stakeholders that have quite different property rights and different powers to change their property rights. The key empirical question is whether stakeholders can cooperate in order to clean up decades of environmental pollution in the "hot spots" known as AOCs. It is the actual construction of the rules of collective and institutional choices that will determine the scale and scope of cooperative action among stakeholders. Before we look at the construction and operation of these rules, in Chapter 5 we will consider some further concerns about institutional rules for RAPs.

5
From Common Property to the Institutional Analysis of Remedial Action Plans

The theory of common property, reviewed in Chapter 4, suggests that institutional arrangements create incentive systems that lead to stakeholders making decisions in relatively predictable ways. The end result of decision making leads to changes in real-world conditions. Different institutional arrangements will have different consequences. Different remedial action plan arrangements will, for example, have different consequences for action in different Areas of Concern.

Over time, stakeholders will develop and amend decision-making arrangements through collective-choice and constitutional processes and will create further kinds of institutional arrangements beyond operational or implementation-type rules. Institutional arrangements consist of laws, regulations, operating practices, and organizational structures that are "stacked" in simple hierarchies. At the lowest level (the operational level), these arrangements or rules consist of procedures or processes that are agreed upon by parties to a decision in order to facilitate decision making. One important set of operational rules is property rights. Operational rules can be changed at the next level in the hierarchy (the collective-choice level), which itself operates within constitutional rules allocating various degrees of authority to stakeholders. Constitutional rules are thus "rules about rules" and have their own processes for amendment. RAPs are institutional-choice arrangements that attempt, in conjunction with other laws, to manage the various uses of the Great Lakes. On "top" of these are constitutional-choice rules of the various state, provincial, and national governments and their component parts (like the EPA and Environment Canada).

Institutional arrangements or rules of various kinds are not fully predictive of real-world outcomes. One key source of variability involves the situation in which the institutional arrangements are developed and applied. For example, we saw in Chapters 1 and 3 that 43 AOCs in the

Great Lakes had been identified as "pollution hot spots," or places where many beneficial uses of the water had been seriously impaired. In these different situations, one would expect different kinds of rules to be developed and applied for restoration purposes.

The characteristics of stakeholders can also make a difference. Human beings vary in their interests, preferences, capabilities to search for (and distill) information, and abilities to learn from experience. Stakeholders also differ with respect to their relative powers. At an operational level, they may possess different property rights – rights that enable them to act in different ways regarding resource decisions and their implementation. At a collective-choice level or constitutional level, certain stakeholders may hold most of the power.

In the case of RAPs, we saw that state and provincial environmental agencies, in consultation with their federal counterparts and related governmental bodies, were given powers to select stakeholders and to approve agendas at the collective-choice level. The International Joint Commission delegated the issues of institutional analysis and design to the American and Canadian governments, which, in turn, "involved" state or provincial governments according to their respective constitutional requirements. In the American case, this involved a new federal statute and a negotiated application of a variety of environmental statutes. In the Canadian case, it involved, in the early years, an executive agreement between Canada and Ontario (the sole province in the Great Lakes basin). In later years, the Canadian government would act alone, although a new executive agreement was drafted in 2001. Granting the powers of institutional design and analysis to one or two government agencies is a "heady mix" – a mix that (as we shall see) most agencies cannot resist. Fortunately, the discretion and power so afforded to agency stakeholders did not manifest itself in matters of narrow bureaucratic self-interest; bureaucracies were engaged in constitution making, as it were, and were quite self-conscious about including and empowering stakeholders.

RAPs were intended to supplement the regulatory regimes of governments in the basin by enlisting stakeholders in the critically polluted areas. The stakeholders were primarily to be other resource users and parties that had some property rights in the resource. The sites, or AOCs, were the situations that were commanding much of the attention of resource regulators. They were usually areas adjacent to urban and industrial centres; they were resources subject to multiple uses and, most frequently, to degradation due to pollution and habitat destruction. In Chapter 1, we reviewed the specific set of impaired beneficial uses for the AOCs; in Chapter 2, we reviewed the current state of pollution in the basin as a whole.

There are four sites of particular concern here: (1) the St. Lawrence River, (2) the Niagara River, (3) the Menominee River, and (4) Hamilton Harbour. These were explored intensively in our original investigations because we needed to examine how the resource situations affected the decision making of stakeholders when they formulated and implemented their respective RAPs.

Each of the sites differed significantly in institutional design. They were selected because of their differing constitutional status, ranging from the most fragmented case (the St. Lawrence) to one of the least fragmented cases (Hamilton Harbour). The St. Lawrence case involved two national governments, one state and one provincial government, a second provincial government that cooperated only on an administrative level, plus two active Aboriginal governments that believed they were (and are) sovereign governments within North America. The Hamilton Harbour case involved only one national and one regional (provincial) government. It was chosen because of its accessibility and familiarity.

The Niagara River case was not quite as complex as was the St. Lawrence River case, but it, too, involved two national and two regional governments. The Menominee River was selected because it forms the boundary between two states (Wisconsin and Michigan) and, consequently, its major constitutional actors are the two state governments and the American government.

We examined RAPs and supplementary scientific and sociological documents for each of the four sites. We conducted in-depth interviews with RAP coordinators and the chairs/presidents of the public advisory groups of stakeholders (the nomenclature of these groups and their roles vary). Below, we present an overview of the four situations.

We also examined some other situations in more depth, although largely through studying secondary source documents. We wanted to look at the two largest AOCs, Detroit and Toronto, plus examples of successful implementations, such as Collingwood (delisted as an AOC in 1994) and Cuyahoga. (Cuyahoga mirrors the Hamilton case in terms of community concern and the relatively simple constitutional framework of one national and one state government.)

Information derived from these four cases was supplemented by information derived from a mail survey sent to RAP coordinators in the 43 AOCs. RAP coordinators are public officials who are either attached to one of the lead stakeholder agencies in an AOC or to contract employees of one of the lead agencies. For the most part, their duties are to facilitate the writing of the plan by teams of public servants (who have the appropriate expertise) and then to develop processes for its coordinated implementation. They are thus central figures in the RAP processes, especially with regard to their knowledge of the stakeholders, their priorities, activities, and interdependencies. At the time of the survey, 39 AOCs had coordinators in place; 32 responded to the survey and follow-up telephone calls.

As the discussion of institutional analysis and design proceeds in this and the next chapter, it becomes evident that some useful information about RAPs, their formation, and their implementation can be culled from government documents, books, and articles about the Great Lakes ecosystems in particular and sustainable development in general (as well as from some such academic works as doctoral dissertations). All of these contribute to a balanced assessment of the dynamics of RAPs.*

In Chapter 6, we present the dominant patterns of behaviour exhibited by agencies as they attempt to change the impairments to the beneficial uses of their respective AOCs. We also examine some important exceptions. Before we do this, however, we need to further develop the rationale behind the institutional rules for managing common property.

The Four Study Sites

Hamilton Harbour
The Hamilton Harbour AOC lies at the western edge of Lake Ontario. Its waters measure 40 square kilometres and are accessible to Lake Ontario by a human-made ship canal (completed in 1830) through a natural

---

*Some of the more important secondary sources were graduate theses on the origins and design of the planning stages of the RAP (Bixby 1985; Strutt 1993; Gunther-Zimmerman 1994; MacKenzie 1996). For other useful volumes on the early experiences with RAPs, see Boyle (1990) and Hartig and Zarull (1992). A number of books and papers that drew evidence from the Great Lakes and RAPs in order to discuss ecosystem and sustainable development policies were also consulted (Caldwell 1988; Hartig and Thomas 1988; Colborn 1990; Hartig et al. 1995; Gebhardt and Lindsey 1996; Hartig et al. 1996). The more useful government documents with regard to discussing progress in developing and implementing RAPs were Eiger and McAvoy (1992); Wayne State University (1994); Hartig and Law (1994); IJC (1997); IJC (1998); and Krantzberg, Ali, and Barnes (1998).

sand bar. The watershed of 500 square kilometres is divided by three main tributaries (Grindstone, Red Hill, and Spencer Creeks).

The watershed has a population of more than 500,000, 95 percent of whom live in urban locations and 60 percent of whom live in the older central city of Hamilton. Manufacturing accounts for 56 percent of employment, including that provided by the largest concentration of iron and steel industries in Canada. These industries, in turn, support the largest commercial port on the Great Lakes, with over 1,000 vessel arrivals per annum.

Some 25 percent of the harbour is reclaimed land, and only 25 percent of the wetlands that existed in 1800 still remain. Seven industrial and four municipal waste point source discharges amount to 27 billion gallons of liquid waste, or 40 percent of the volume of harbour waters. Water quality conditions include elevated levels of phosphorous, ammonia, heavy metals, and PAHs. According to IJC criteria, the harbour has nine impaired beneficial uses, one of which (beach closures) was restored in 1995.

Menominee River
This AOC encompasses the lower 5 kilometres of the river that forms the boundary between Wisconsin and Michigan. It consists of the portion from the upper Scott Paper Company Dam (Wisconsin), flowing between the two sites of Marriette and Menominee, to the river mouth on Green Bay. Also, some 5 kilometres of the shoreline on either side of the mouth are included in the AOC, along with six river islands, which are part of Wisconsin. The watershed drains some 103,600 square kilometres and the river itself flows 184 kilometres.

The AOC has a population of approximately 25,000, with some 12,000 in Marriette and 10,000 in Menominee. Early industrial development, especially in lumber, occurred around the mouth of the river, taking advantage of waterway transportation. Industrial land use remains prevalent along the riverfront. There are six major point sources of pollution, resulting in adverse water quality. These include contaminated sediments (especially arsenic and PAHs) and periodic elevated loadings of BOD (biological oxygen-demanding wastes that need oxygen as they decompose), suspended solids, phosphorous, and heavy metals. Upstream navigation is now limited by two Scott Paper Company dams.

The river mouth area is distinguished by the 20-acre Seagull Bar Natural Area and the underdeveloped Green Island. These natural habitats contrast with the contoured lower reaches of the river, where much of the wetlands were reclaimed. This AOC has six impaired beneficial uses.

Niagara River

This river is a connecting channel between Lakes Erie and Ontario, and it contributes over 80 percent of the annual flows into Lake Ontario. This AOC extends the entire 58 kilometres of the river. It includes the Welland River watershed on the Canadian side a nd seven small watersheds on the American side. Most of Buffalo is excluded from the AOC, as it drains into the Buffalo River. (The Buffalo River, however, is a separate AOC.) Due to a history of conflicts over toxic pollution in the Niagara River, the Canadian and American sides have developed their own separate RAPs. However, an international advisory committee shares information and facilitates cooperation. In addition, the four senior governments (of the United States, New York, Canada, and Ontario) are parties to the separate Niagara River Toxics Management Plan (NRTMP), which runs parallel to the RAP. The NRTMP focuses on measuring 18 primary toxics both upstream and downstream.

Approximately 400,000 persons live in the AOCs, with the largest concentrations being adjacent to Niagara Falls and the City of Buffalo (North and South). The development of hydropower in the early twentieth century attracted major industrial developments, particularly on the American side, and included major chemical, steel, and heavy manufacturing plants. Currently, over 50 percent of the flow of the river is diverted for hydropower. The Canadian AOC consists largely of small towns supported by agriculture and tourism. The most southerly ports of the US AOC contain the remnants of the once-thriving steel port of Buffalo.

Developments along the river have resulted in "dramatic" losses of wetlands and shore habitats for fish and wildlife (New York State, Department of Environmental Conservation 1994, 4-99). Contaminated sediments, inactive hazardous waste sites (e.g., Love Canal), and discharges from 26 American and six Canadian point sources contribute to highly elevated levels of major toxic organics and inorganics. The major Canadian source, contaminated sediments in the Welland River, was removed by dredging in 1996. Delisting of the AOCs (ten use impairments in Canada; seven in the United States) remains contingent upon success on the American side.

St. Lawrence River

The headwaters of the St. Lawrence River are the site of the two most easterly AOCs. The Cornwall AOC, which extends from the Moses Saunders Power Dam at Cornwall to the Beauharnois Power Dam near Montreal, is a Canadian AOC based on the northern shoreline of the river. The Massena AOC, which runs on the southern shore from Massena, New York, to the River à la Guerre in Quebec, is an American AOC. The Canadian AOC has five watersheds, including three in Quebec, while the American AOC drains three small watersheds. Both AOCs include lands belonging to the Mohawk Nation at Akwesasne. Neither the Akwesasne (who withdrew) nor the Province of Quebec (which never joined) are official participants in the two RAPs. Early attempts to establish a single international AOC floundered. However, both RAPs use a "transboundary impacts" use-impairment indicator in addition to the 14 IJC use-impairment indicators.

The population of the two AOCs is approximately 190,000, including some 11,000 Mohawks. The largest town is Cornwall, Ontario, with 70,000 people. Both AOCs are primarily agricultural, but the development of the seaway and associated hydropower (four dams) attracted industrial plants such as Alcoa, Reynolds Metals, and General Motors Central Foundry (on the American side) and Domtar Fine Papers and Courtaulds Fibres and Films (on the Canadian side). (The Courtaulds plants are now closed.) Cornwall itself grew, first, due to an extensive textile industry (now defunct) and, second, due to the development of the St. Lawrence Seaway. Some 3,000 vessel transits a year pass through the Montreal-Lake Ontario section of the seaway.

The rocky shorelines of the St. Lawrence, unlike its tributaries, do not favour fish and wildlife habitat. However, the improvement of Lake St. Francis and the construction of the seaway had major impacts on fisheries habitat. Major sports, commercial, and subsistence fisheries remain. There are ten point sources of pollution in the Canadian AOC that are contributing elevated levels of bacteria, BOD, suspended solids, mercury (around Domtar), and zinc (around CIL). There are 23 small municipal point sources of pollution and 14 industrial point sources in the American AOC. Elevated levels of PCBs, dioxin, mirex, PAHs, and mercury are associated with wastewater discharges and hazardous waste sites owned by Alcoa, Reynolds, and General Motors. There are seven use impairments in the Massena RAP and 12 in the Cornwall RAP. The transboundary indicator is considered impaired; it is believed that this is caused by the transport of PCBs, metals, contaminated sediments, and nutrients to downstream Canadian sites.

Framework for Institutional Analysis

We posed, on the basis of the theory of common-property resources, that outcomes (such as the pollution or remediation of an AOC) were the product of situational variables (such as the ecosystems of the area) and institutional variables that affected resource decision making and implementation.

The basic frameworks (I and II) for institutional analysis were introduced in Chapter 4 (see Figures 4.4 and 4.5 on pages 71-72). Basic framework II essentially amplifies basic framework I. We will now extend our discussion of these three classes of variables (situations, structures, and outcomes) and explain what they mean and how they are likely to be found in our empirical investigation.

Basic and Amplified Frameworks
If our analysis is correct so far, then RAPs adopted and implemented in the AOCs should have an impact on their ecosystems. One major indicator of these impacts is restoration of the 14 impaired beneficial uses. The goal of the government parties is to restore the AOCs by around the year 2010. The term used by the International Joint Commission to refer to this restoration is "delisting." So far, one AOC (Collingwood) is delisted.

Environmental science is often uncertain, vague, and ambiguous, and this affects the selection of valid indicators of impairment, delisting, and restoration improvements. For example, impaired use number 11 is the presence of fish tumours or other fish deformities. The inclusion of this impaired use as one of the 14 impaired uses can be justified on both intrinsic grounds (it indicates direct ecosystem degradation) and instrumental grounds (it can negatively affect human health through extensive consumption). However, for evidentiary reasons, both justifications are uncertain. Further, the listing guideline indicates that an area should be listed "when the incidence rates of fish tumours or other deformities exceed rates at unimpacted control sites or when survey data confirm the presence of neoplastic or preneoplastic liver tumours." (The delisting guideline indicates the converse.) The rationale for this particular guideline is that "it is consistent with expert opinion on tumours [and] it acknowledges background incidence rates" (IJC 1991b). Thus, at any time, it can be challenged by new science. Appendix A lists the IJC criteria and their rationales.

Even the measurement of a guideline can be challenged for its validity. For example, the incidence rule of fish tumours in bullheads or suckers in the Black River (Ohio) AOC is measured in two ways: (1) no neoplastic liver tumours in a minimum sample of 25 brown bullheads (≥ 2 years old); and (2) the incidence rule of skin and lip tumours must be less than the incidence rate in a control site. A finding of 150 control-site and 130 contaminated-site fish would be needed to verify a 5 percent difference (Hartig et al. 1994, 345). Appendix B lists measurement examples of each impaired use.

These kinds of uncertainties suggest some imprecision with regard to indicators or measures of a valid ecosystem outcome. The range of uncertainty is reduced, however, if we are less interested in precision than we are in estimating whether some degree of improvement (e.g., large, moderate, or small) has taken place with regard to an impaired use. Again, while the selection of the 14 impaired uses is debatable, as a whole they do measure some degree of ecosystem degradation or non-sustainability. And, concomitantly, a change in the right direction for one of these impairments can be estimated, albeit imprecisely, when seeking some valid proxies of ecosystem sustainability.

In addition to these so-called "objective" indicators of remediation, we also used "subjective" estimates: improved beneficial uses, improved water pollution problems, and improved beneficial uses due to the RAP that was provided by RAP coordinators. These were supplemented by estimates provided by secondary sources for particular AOCs as well as by case study evidence.

We also classified a number of implementation measures as outcome measures, although, strictly speaking, we would call them operational rules. This is because RAPs tended to produce some newer implementation processes, and we wished to uncover these as part of the results of the RAP deliberative processes (collective-choice arrangements). RAPs also used established processes of implementation where one or more of the governmental and private organizations were listed as stakeholders; we wished to disentangle these conventional processes from the newer ones adopted by each RAP. In general, we wanted to see which agencies were involved in implementation, and what agency might or might not be in a lead or control position with regard to the operational rules.

It is also important to note that the rules for implementation could well differ from the rules for the planning of remedial actions. First, stakeholders might well prefer to delegate projects to a small set of implementers in order to ensure speedier decisions. In other words, the transaction costs of implementation might well be lowered by delegating planning actions to one or more agencies. We would anticipate that delegation to one single agency would occur only when the planning action was a definable activity; a multiple-use resource like an AOC would probably require the coordinated actions of a number of organizations.

Implementation of planned actions necessarily follows the planning phase, but there is some overlap between the two sequences. First, extensive plans often include implementation strategies and activities and indicate the parties responsible for implementing the strategies and activities. The RAP Stage 2 reports all required an implementation section for acceptance by the IJC.

Second, RAPs were not beginning remediation de novo. Substantial work on remediation, regulation, and prevention had occurred and was occurring in AOCs prior to the RAP process. The strategies and activities of implementation would need to be integrated with ongoing efforts to avoid unnecessary interruptions of current programs.

Third, we know from many decades of work in public administration that the implementation stage is never bereft of planning and policy choices. RAPs are being revised and altered as the implementing parties make choices about priorities and methods of implementation. These need not be inconsistent with the overall plan. But values and facts are intermingled as choices must be made by implementers (V. Ostrom 1974).

Given these implementation issues, one might depict a number of alternative institutional arrangements for implementation. Three kinds of arrangements suggest themselves from the theory and practice of management across interdependent organizations (Sproule-Jones 2000).

The first arrangement is referred to as pooled coupling, whereby organizations are assigned specific responsibilities in the plan or in a subset of the plan (for particular activities). Each organization can proceed with implementation, and while complete implementation requires the effective success of all parties, any one party can proceed even if others are delayed or are postponing their assigned responsibilities. A second arrangement is referred to as sequential coupling, whereby agency C can only implement if agency B has completed its tasks and if, in turn, agency A has successfully implemented its tasks. In some literature (e.g., Thompson 1969) these are referred to as long-linked technologies. The third arrangement is referred to as reciprocal coupling, whereby agency A needs the cooperation of B and C in order to effect a second or recurring phase of its tasks. There is a continual exchange of activities on the part of responsible organizations.

All three arrangements will carry with them differing sets of transaction costs, depending on the complexity of the tasks and the number of coupled organizations. One would also expect that the reciprocal form of coupling would require continuous monitoring and coordination as one break in the linkages could negate the whole set of interlinked activities. The sequential arrangements would also be subject to the threat of breakdowns in the coupling linkages.

In our investigation of RAPs, we will be looking for the various forms of coupling that take place between coordinated agencies and examining the differing rules put in place to deal with such coupling difficulties that may emerge.

In short, our analysis of outcomes will include substantive outcomes in relation to the remediation of the 14 impaired beneficial uses of the AOCs. These substantive outcomes will be indicated through scientific reports and the perceptions of key decision makers. Our analysis of outcomes will also include process outcomes; namely, the rules put in place to implement RAPs across a variety of diverse but interdependent implementers.

Structure

Structure refers to the collective-choice arrangements under which RAPs were formulated. What were the processes or rules through which different interests were or were not included in the decision making? And how were these interests aggregated or weighted by their relative powers in the process? Both these questions are central to decision making at any collective-choice level – federal, provincial/state, local, or watershed/AOC.

We want to know something about the exit/entry rules to the collective-choice process. How are stakeholders selected? How can they join and leave the groupings? Theory would suggest that the more inclusive the membership of stakeholders in the process, the more efficient the ultimate decisions. This is because stakeholders can articulate their concerns directly and make adjustments, trade-offs, and compromises with regard to collective needs (Sproule-Jones and Richards 1984). Indirect representation by others is more likely to introduce errors when calculating the intensity of any stakeholder's concerns, and the marginal value of these concerns, against other values articulated in the process. Even more extreme, zero consideration may mean that a particular set of interests is excluded. So, for example, it is best to include recreational fishers in the decision-making process if their interests are to be articulated rather than guessing or ignoring their political concerns and relevance to a multiple-use harbour, river mouth, or connecting channel. Negotiation and bargaining among stakeholders is a central feature of resource decision making for multiple uses. Stakeholders may be seeking to gain and sustain rents (or a non-competitive advantage over rivals) for the use of the resource. On the other hand, they could be advancing efficiency if their property rights are well defined (Coase 1960) – a situation that seems unlikely in the RAP case.

Comparably, the interests, once represented, need to be aggregated into a final product. One model for doing this might be the following: a lead bureaucracy is named by the relevant constitutional rules (e.g., state officials in consultation with Environmental Protection Agency officials); this bureaucracy acts to organize expert opinions on how best to technically remove the use impairments and (in turn) to bring these opinions and options to the other stakeholders for choice and amendment. A different model might convene a forum of stakeholders in order to define goals for remediation; these goals could then be used to calibrate and prioritize the range of cleanup strategies that could be developed in dialogue with technical experts. In either case, the collective-choice rules for aggregation could vary from unanimous consent (or consensus) among all stakeholders to domination by one (or a few) stakeholders under an implicit rule (of constitutional authorities) that calls for delegating formation to a small set of interests.

We know from theoretical work that increasing the rules of aggregation may increase the transaction costs of decision making as different interests must be included in any winning coalition. We also know that these transaction costs can be increased exponentially by the strategic and opportunistic behaviour of stakeholders as they exploit their "veto-like" positions. If all stakeholders agree that a particularly important industrial plant must be a partner to a collective decision, then the owners of that plant could indulge in opportunistic behaviour because of the incentives afforded by the partnership rule. Again, the transaction costs of negotiation are increased by such opportunistic costs.

Transaction costs will fall dramatically with decision making that is dominated by a few stakeholders and legitimized by decision-making rules that grant final say to a few interests. The bargaining will be reduced in time and effort, as will the incentives for opportunism. However, the probability of errors in weighing and balancing multiple users and interests will increase. Stakeholders may well resort to destructive strategies if their interests are ignored; turn to the constitutional-choice level to see if decisions can be overturned; or simply opt out, leaving thehypothetical RAP just a wordy document that is ignored in practice by disgruntled stakeholders.

Figure 5.1 is a graphic representation of the major dynamics involved in the aggregation of user interests. It is derived from Buchanan and Tullock (1962) and Sproule-Jones (1974), with some changes in nomenclature. It should be noted that the costs of collective action are zero in terms of domination of one interest over another when consensual decision making is adopted. It comes, however, with higher transactions and opportunistic costs. As well, domination costs can escalate when decision making is limited.

It may be suggested that one of the acts of successful coordination and leadership is to minimize domination costs but, at the same time,

Figure 5.1

Aggregation of user interests

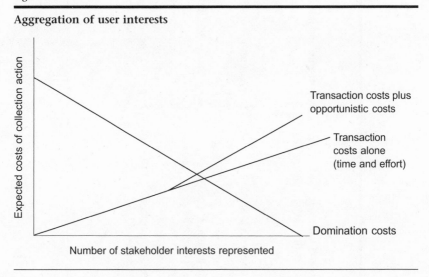

Number of stakeholder interests represented

to work to minimize transaction costs and to preclude or foreclose on strategic opportunism. On the other hand, coordination and leadership may not adopt such a strategy unless the situation calls for repeated plays of the game among interests that trust each other (E. Ostrom, Gardner, and Walker 1994).

We noted earlier that the IJC delegated to respective governments the establishment of articulation and aggregation rules for each AOC. A lead bureaucracy took this responsibility in each AOC under constitutional decisions worked out between the respective federal and state/provincial governments. There were no understandings for institutional design other than the involvement of stakeholders in a public participation process. This placed the bureaucracies in question in a situation in which the rules devised could vary widely, from rules encouraging consensus among inclusive groups of stakeholders to rules minimizing transaction costs due to the informational impact of selected stakeholders. Bureaucracies were thus placed in a monopoly, or single-sovereign, position (V. Ostrom 1989).

Monopolistic behaviour is best characterized as discretionary behaviour. Within the context of markets for private goods, this discretion can be used by a monopolist to generate monopoly profits through restricting the supply of goods. Within the context of public goods, it can be used by a monopoly bureau to increase service production levels or to generate "organizational slack" for productive or non-productive

Figure 5.2

Working conceptual framework

Situational variables *(Pre-conditions)*		Structural variables *(Constitutional and institutional arrangements)*	Outcome variables
Ecosystem conditions	Community characteristics	Aggregation and articulation	Implementation and impacts
1 Number of impaired beneficial uses 2 Three worst pollution problems 3 Other	1 Population 2 Income 3 Municipal fragmentation 4 School boards 5 Other special boards 6 Pluralist index 7 External support for research, etc. 8 Education	1 Selection of stakeholder 2 Rules for representation 3 Rules for aggregation 4 Inclusivity of rules 5 Role of federal/provincial agencies 6 Role of writing team 7 Membership turnover	Novel/Single/Multiple Lead/Control Remedies (sanctions) Monitoring/Indicators Improved uses Improved problems

purposes (Niskanen 1971; Breton and Wintrobe 1979; Dunleavy 1991). We should thus expect a variety of responses from monopolistic bureaucracies charged with constructing the institutional design of RAPs.

In terms of collecting evidence, we must look for information about the different roles of government agencies and of technical advisory committees (sometimes called writing teams) in the formulation of RAPs. We must also look for information about the correlative roles of other stakeholders in establishing one or more of the goals, objectives, and strategies in RAPs.

Situations
The situational differences between AOCs are best summarized by the number and degree of impaired beneficial uses designated by the IJC (see Chapter 1). Each AOC differs from others in terms of its important ecosystems, its particular history of research exploitations, its pertinent stakeholders, and the communities around its shores. We will examine further evidence about the worst pollution problems in the AOCs as well as a range of other variables pertaining to the ecosystem and communities. Whether they have a negative or positive impact on RAP formulation and implementation, the size, diversity, institutional fragmentation, and scientific resources available to each community are assessed. We capture all of these measurement concerns in Figure 5.2, which is an expanded framework that we refer to as a working conceptual framework.

Conclusion
We are now in a position to distill the results of our research into a conceptual framework. In this kind of exercise, we do not propose to develop a precise model of institutional analysis and to verify its causal connections with quantitative data; rather, we are concerned with revealing the central propositions of institutional analysis that explain the RAP processes in particular and large-scale, multiple-use processes in general. In essence, we ask the question, "Under what institutional arrangements can holders of different bundles of property rights agree to remediate their immediate aquatic environment?" The answer to this question should give future decision makers some good information about the structure of planning and implementation rules as well as about their linkages to effective policy outcomes. In the short term, we can begin to assess the performance of our institutional rules for the restoration of the Great Lakes.

6
Patterns of Behaviour

The theory of common property isolates the nature of rules (or, as we might prefer, institutions) as an important set of causal factors in explaining environmental and resource management behaviour. These rules/institutions may be analyzed and divided into rules at the constitutional, collective-choice, and operational levels of decision making. At a collective-choice level, stakeholders in polluted Areas of Concern on the Great Lakes have, and are devising, restorative plans and implementing them in coordination with each other. The end product should be, we hope, improved environmental outcomes or, more specifically, restored beneficial uses.

Remedial action plans have been, and are still being, formulated and implemented. Some began as early as 1985; others are still to begin. AOCs are at various stages in the RAP process. This enables us to look at RAP progress as it proceeds through its various stages of problem definition, recommendations for restoration, and delisting. Our survey work was conducted when 4 percent of RAPs were still to begin, 48 percent had completed Stage 1 (problem definition), 44 percent had completed Stage 2 (recommendations), and 4 percent had completed Stage 3 (completed implementation). Our case study work includes sites at the Stage 1 and Stage 2 levels. The key questions are, "What is really happening?" and "How can we explain these happenings?"

Outcomes: Substantive

If we look at the end product of RAP processes, then we discover that RAPs have improved a few uses. RAP coordinators (the public official who facilitates the coordination of RAP implementation in each AOC) report that a few improvements have occurred in 68 percent of the AOCs, particularly in the areas of water chemistry and biota. The Ontario government calculates, again based upon RAP coordinators' reports, that

51 percent of restoration targets in their 12 AOCs have been attained (Krantzberg 1998). Comparable figures exist, albeit at earlier dates, for American sites (Hartig and Law 1994). So far, so good.

However, if we inquire further into the particular improvements that are evident, then we discover that bacterial pollution and lowered point source discharges of toxins (like PCBs) are occurring in roughly two-fifths of the AOCs. Comparably, improvements in biotic problems and human uses of sites are occurring in roughly 30 percent of the AOCs. But the pollution reductions and improvements to beneficial uses are due to programs that existed before RAPs were implemented. Regulations on industrial discharges and municipal sewage treatment plants were largely in place in the mid-1980s, and improvements are reported to be associated with these regulations and their refinement. One RAP coordinator summarizes the RAP program as follows: "Implementation is an ongoing process because we have existing programs that came before the RAP process. The RAP process just coordinates the individually mandated existing programs that are driving the environmental control programs."*

Case study evidence reinforces this mixed conclusion. A recent (1999) update of the Hamilton Harbour RAP reports significant progress in fish and wildlife habitat restoration (made possible through new RAP programs and resources) but only modest improvements in some water quality indicators (Knox 1999). The St. Lawrence (Massena) RAP coordinator reports that "all of the seriously contaminated site cleanups were already underway under federal and state superfund programs or under EPA unilateral cleanup orders ... The RAP was not leading the way in defining problems or identifying the programs required to clean them up ... [But] we have not seriously looked at the non-contamination related issues such as wetland restoration or fish and wildlife habitat restoration."

Similarly, the Menominee (Wisconsin) coordinator comments that "the major environmental problems are being addressed through other [than RAP] enforcement programs ... the coal tar site, the paint sludge site and the arsenic site. Each year we have a bit of Great Lakes Pollution money that can be used for specific issues that are not covered by existing programs." The apt conclusion seems to be that some progress is being made in remediation but that this may, in large part, be due to

---

*This and subsequent quotes are taken from either the mail survey of RAP coordinators in the summer of 1996 or from personal interviews conducted in 1997 with coordinators and RAP participants.

pre-RAP programs. It could be concluded that the agencies are choosing to pursue their traditional concerns and that the incentives provided by RAPs (some extra resources and a coordinated response by multiple stakeholders) are insufficient to change or speed up these continuing programs.

There is another institutional incentive that reinforces traditional programs rather than newer RAP efforts. While AOCs have their own combination of impaired beneficial uses, RAPs could (1) define their own set of priority problems that might be linked to beneficial use impairments and (2) identify their own particular set of "stressors" that need to be removed in order to improve their own ecological situation. This gave the implementers of each RAP (who are often collaborators on the authorship of the plans) sufficient discretion to deal with what they perceived to be their major problems. Our survey revealed 18 different priority problems across the AOCs. The cases revealed similar patterns. For example, in the St. Lawrence case, the two RAPs could identify a range of restrictions on fish consumption, but the linkages to sources of pollution are acknowledged as uncertain. In the face of this uncertainty, the American Massena RAP continued to focus on remediating PCB pollution from sites and discharges of three industrial plants (Alcoa, Reynolds Metal Aluminum, and General Motors Power Train Division). On the Canadian side, the Cornwall RAP continued to focus on inorganics (like mercury and zinc) associated with the two industrial plants owned by Domtar Fine Papers and Courtaulds (the latter now closed), and the industrial discharges emitted by the Cornwall Sewage Treatment Plant.

These sources of pollution on both sides of the St. Lawrence might well be linked to fish consumption advisories, although no more than, say, the physical construction of the St. Lawrence Seaway and the associated destruction of wetlands. RAPs could and did make choices regarding priority problems. These tended to reflect the ongoing priorities of the regulatory agencies rather than a new mix made possible through a multiple stakeholder process. This point is further explored below.

Contrasting Differences

RAPs are modestly successful in terms of outcomes, although much of the credit for this is due to pre-existing programs rather than to RAPs themselves. But we are talking of the overall pattern. In contrast, we can highlight two examples of the worst and best situations, both of which point to the importance of institutional design in facilitating successful or failed outcomes.

Port Hope, on Lake Ontario, is a small harbour on the Canadian side that has one single use impairment, that of contaminated sediments (Hartig and Law 1994, 164-66; Canada-Ontario RAP Steering Committee 1995, 58-60). Approximately 90,000 cubic metres of low-level radioactive wastes (plus some heavy metals) exist in the turning basin and west slip areas due to the disposal of wastes from the refining and processing of uranium and radium during the 1930s and 1940s. A multistakeholder public advisory committee (PAC) working as an advisor to a RAP writing team (a conventional model) helped to establish a Stage 1 report outlining goals, problems, and options. The Stage 2 report on implementation is still to be completed because the disposal options for the contaminated sediment is controlled by a different Canadian government agency. This agency, Natural Resources Canada, has its own siting task force and community liaison group whose purpose is to find an appropriate remediation option. This is a case in which local environmental conditions (waste disposal siting), allied with an unambiguous legislative mandate given to the Atomic Energy Board (regulated through Natural Resources Canada), made the local RAP process peripheral to the restoration of the AOC. A key stakeholder was not included in the collective decision making, and this stakeholder is prepared to play a non-collaborative game in order to satisfy its monopoly position over nuclear waste disposal. Over time, it developed its own plan, sought special appropriations from the Canadian government, and is now beginning to implement it.

In contrast, Collingwood Harbour on Lake Huron was delisted as an AOC in 1994. The local contributions to improved beneficial uses were remediated through sewage treatment plant optimization, dredging and capping of shipyard slips, and non-point source controls. The public advisory committee that, with a government writing team, formulated the RAP still exists to promote and to steer aspects of the plan not directly linked to use impairments. These include habitat enhancement, water conservation, and public outreach programs (Hartig and Law 1994, 80-83; Canada-Ontario RAP Steering Committee 1995, 31-34). In this second case, unlike the Port Hope case, local environmental conditions were amenable to an available set of solutions. No stakeholder monopoly or opportunism could jeopardize a fairly rapid implementation (six years from the Stage 1 start to the Stage 3 submission that led to delisting).

Before we examine the institutional rules used to design RAPs, we will look at some of the processes that exist for implementing RAP recommendations.

Implementation Outcomes

The implementation strategies and methods developed by RAPs were and are a product of the RAP planning processes. We are thus calling them outcomes, even though, strictly speaking, they are processes rather than products of institutional rules (in other words, remediation does not occur in the AOCs because these implementation strategies were adopted). Nevertheless, we are calling them outcomes.

The favoured method of implementing the RAP recommendations is referred to as "pooled coupling." This method relies on multiple organizations for implementation, but each organization is a lead or designated agency for different activities. Some 47 percent of RAPs use this as their major strategy. Some (8 percent) attempt to use "sequential" and "reciprocal" coupling, which involves joint control of all activities, as the predominant method. Roughly 42 percent were attempting to implement RAP recommendations under more hierarchical arrangements, whereby one agency was designated as in control of implementation. An apt illustration of the two major methods of implementation occurs in the St. Lawrence River. On the Canadian side, the RAP coordinator comments: "We have identified who the potential partners are for each recommendation and it varies depending on the recommendation." Across the river, the RAP coordinator notes that "the DEC [New York Department of Environmental Conservation] is in charge of implementation under various state and federal programs."

When we examine our AOC cases in detail, we see that, in practice, the coupling arrangements tend to be found in combinations; that is, there may be a dominant form of coupling – that of pooled coupling – but the other forms may be used to a lesser extent in the same AOC. Hamilton Harbour provides good examples.

In pooled coupling delivery systems, different organizational units are each responsible for one or more of the activities that comprise a program. In our Hamilton Harbour case, some 14 organizational units are each responsible for implementing one or more of the RAP recommendations for restoration of the harbour ecosystem. If one unit fails or stalls its implementation, then the other units can still proceed. A CDN$2 million fish and bird habitat project was completed in 1999 in advance of major improvements in water quality parameters associated with the municipal sewage treatment plants.

Sequential coupling occurs when the work flows take place in long linked chains and when one activity cannot proceed before others occur. In our Hamilton Harbour case, for example, an $8 million marsh restoration project was organized (1995-2000) by the sequential coupling of public and non-profit organizations. The Royal Botanical Gardens (RBG) bought cattail seeds, which were distributed by the Bay Area Restoration Council (BARC, a non-profit monitoring and educational organization) to some 140 Grade 7 classrooms under agreements with three school boards. The young, grown cattails were collected by volunteers working with BARC, which stored them at the RBG greenhouses for eventual planting by the Friends of Cootes Paradise. The latter group is a non-profit organization consisting of 70 volunteers; these people are organized by a McMaster University professor to plant and extend the marshland of the tributary to the harbour known as Cootes Paradise.

Reciprocal coupling consists of exchange relationships among units. Unit A produces an activity for Unit B, which, in turn, may pass it back to Unit A for completion. It could also involve the simultaneous coupling of units. The reciprocity may involve two or many units, thereby increasing the possibilities of uncoupling as the number of units increases. One Hamilton Harbour example of reciprocal breakdown involves the contaminated sediment dredging project known as Randles Reef, in which 20,000 cubic metres are scheduled to be removed, treated, and disposed of by four units under a jointly financed partnership. The technical proposals have been successfully blocked for ten years by one partner, Stelco (a steel company), despite the enthusiastic endorsement of BARC, the federal and provincial ministers of the environment, and the three other units engaged in the coupling (Environment Canada, Ontario Ministry of the Environment, and the Hamilton Harbour Commissioners, respectively). The irony is that the contamination probably occurred after a coal-tar spill by the steel company, possibly before but probably after the regulatory legislation was proclaimed in 1975. The reciprocal coupling will probably have to change to pooled coupling if the project is to be completed without Stelco.

This evidence suggests that each AOC may have adapted its implementation strategies to the particular configurations of stakeholders represented in the particular remediation situations at hand. Predictably, not all are successful or appropriate. We need to unravel this issue.

Structural Factors

Representation

Our conceptual framework suggests that the various outcomes of the RAP processes are due, in part, to how stakeholders are represented and how their interests are aggregated into a planning document. At the level of constitutional choice with regard to establishing collective governance systems for AOCs, we have noted that the IJC allows the respective federal and state/provincial environment agencies to define the rules for the articulation and aggregation of stakeholder interests and for the implementation of a RAP. The discretion so afforded is wide, and a prevailing pattern of "business as usual" pervades the planning and implementation of RAPs. We have also noted that exceptions to this system can and do exist.

At an institutional level, state/provincial and/or federal environmental agencies frequently selected and approved the stakeholders that were to participate at the planning stages (RAP Stages 1 and 2). This resulted in an unrepresentative configuration of interests in many AOCs. Our surveys suggest that most stakeholder committees included federal, state/ provincial, and municipal representatives, plus representatives of industry, farming groups, environmental groups, universities, and "citizens at large." Of the major users, recreational, shipping, and human health groups were frequently not included, and Aboriginals were included in only one of our AOCs. In the St. Lawrence RAPs, the Mohawks of Akwesasne withdrew their participation. According to their representative:

> The Mohawk governments agreed to participate and promote the RAP if certain conditions were met. One of these conditions was that the Governments of Akwesasne were to be recognized as an equal partner. On the Canadian side this was recognized and a Mohawk government representative sat on the RAP Team [the scientific technical team that wrote the Stage 1 document]. On the American side, the Mohawk government was invited only as a member of the public. We refused membership ... By 1990, the MASH [Mohawks of Akwesasne for Safe Health] representatives did not see any respect from the [Canadian] public advisory committee. The major problem was there was no democratic or representative process. We came to see the PAC as a way the Canadian government could rubber stamp their research agenda.

This view was confirmed by the two RAP coordinators for the St. Lawrence.

In the Niagara River (NY) AOC the Tonawanda reservation was not invited to participate because it owned no shoreline property. Again, this perception of the status of Aboriginal peoples was different from their perception.

It is worth noting that the New York Department of Environmental Conservation (DEC) rejected the concept of a single international RAP for both the St. Lawrence and the Niagara Rivers. They wished to "avoid these bureaucratic problems where you have different political systems and different goals ... [Also] ... our regulations were much firmer than the Canadians'." The Menominee River RAP experienced some of the high transaction costs of agreement between Wisconsin and Michigan that were so feared by the New York DEC. Ostensibly, Wisconsin was the lead state for the river because it only had five AOCs while Michigan had 14. But the Wisconsin Department of Natural Resources felt that the Michigan Department of Natural Resources was lax in its contributions to the Stage 1 report and that it was generally lax regarding regulatory enforcements. The RAP coordinator stated: "We finally agreed ... that Michigan should comment on the contamination sites on their side of the river and Wisconsin was to do the same ... As long as there is not a major conflict, we were not to interfere with remediation recommendations at those sites." In practice, this is virtually the same as having two separate RAPs. The federal Environmental Protection Agency appears loath to facilitate an integrated approach, focusing instead on applying its enforcement programs to identified problems in the river. In essence, the presence of two or more sovereign governments has the potential to destabilize resource management decision making.

In sum, most interests are included in most RAP public advisory committees (PACs). This is easier in the larger PACs, and it should be remembered that over 48 percent of RAPs have PACs that are made up of 24 or more persons. Significantly, prior to their appointments, almost 60 percent of the committee members had attended PAC meetings or contacted the RAP coordinator about volunteering. Thus most RAPs were not designed to be exclusive.

Agendas

As important as the representative character of the stakeholders' committee is, the capacity of these committees to set common goals and to establish common recommendations resulting in the achivement of these goals is just as critical. At this stage, the institutional designers exercise major controls over the process. In approximately four-fifths of RAPs, the relevant state/provincial environmental agency and the seconded public servants on the writing teams set the goals and made recommendations for remediation. The other stakeholders were invited to provide input for amendments.

The Niagara River (US) RAP is fairly typical. Their PAC chair comments: "What would happen is that the DEC [Department of Environmental Conservation] would prepare a draft of a chapter. It would come to the committee for review. We would send it back, it would be revised and come back to us." In contrast, the Menominee chair believes that their PAC was "very influential. Particularly in the Stage 1, but also in the [Stage 2]. The Citizens Advisory Committee really went page by page through the document." Consequently, in this case, there was often disagreement between the Citizens Advisory Committee and the Wisconsin Department of Natural Resources. For example, Stage 2 was a focus of conflict. The coordinator comments: "The CAC threatened to remove all references to [itself] if their demands were not met. Such as the Menominee Fisheries Plan. The fisheries plan was controversial. There were references to the removal of dams and the installation of fish ladders. The CAC opposed this. Many times the CAC looked out for the best interest of the local economy, but did not necessarily work in the best interest of the River."

Two exceptions to the prevailing pattern of local groups acting as advisors to state/provincial administrators were Hamilton Harbour (Ontario) and Cuyahoga River (Ohio). In the Hamilton case, a group of 50 stakeholders formulated goals for Stage 1 by March 1988 and draft recommendations for Stage 2 by June 1989. Stage 2 was finally approved in 1992. The writing team organized itself as technical support for the local endeavours. This model was not copied in other Ontario AOCs because the Ontario Ministry of the Environment realized that this could result in its losing control over the RAP process to local stakeholder interests. In Cuyahoga, the Ohio Environmental Protection Agency designed a Coordinating Committee (CCC) of 35 stakeholders drawn from multiple interests and sectors and gave it express authority to develop Stage 1 and Stage 2 documents.

We thus have a variety of examples of stakeholders acting as consulting organizations to state/provincial agencies in order to identify the problems and solutions necessary for remediation. We have fewer examples that involve the stakeholders actually identifying the problems and solutions, thus making the necessary trade-offs and compromises to accommodate efficient multiple uses. A suggestive indicator of the importance of decentralized decision making (by stakeholders rather than by agencies with stakeholder advice) is the perception that the RAP has improved beneficial uses due to its use of inclusive decision making and due to collaboration between the stakeholders and the writing team (both were statistically significant, at 99 percent probability.) The top-down model of most RAPs is seen as less effective.

Situations

As we saw in Chapter 1, AOCs differed in the number of impaired beneficial uses, as defined by the IJC. Obviously, the more impairments, and the greater their severity, the more difficult it is to develop successful remediation. However, there are other situational factors that could influence the RAP processes and their success. We have seen how AOCs that span international and state lines have decision-making difficulties and must develop parallel processes in order to account for shared jurisdictions. Fortunately, almost 90 percent of the AOCs are contained within a single national and a single state/provincial situation, and over 66 percent of them contain no Aboriginal reserves/reservations. Again, some 70 percent of the AOCs are situated within no more than six municipal government jurisdictions. Institutional fragmentation of government is not an obstacle for most RAPs, and it is not correlated with organizational processes for implementation.

However, the size of an AOC, in terms of population, may be an issue as it signals greater urbanization and greater impairments of the local environments. Eighteen percent of the AOCs have populations greater than 500,000 people; two, the Detroit River and the Metro Toronto and Region RAPs, have populations in excess of 4 million. As of 1998, the Detroit River RAP had completed only 5 percent of the remedial actions (identified in its Stage 2 report) for its eight impaired beneficial uses (Krantzberg 1998, 38). Metro Toronto is roughly comparable, although the extent of toxic contamination in discharges and sediments and the scale of the loadings are substantially less than they are in the Detroit "connecting channel." Metro Toronto has completed only 20 percent of its Stage 2 activities (Krantzberg 1998, 53).

The Detroit River has a long history of degradation manifested in contaminated sediments and destroyed wetlands (IJC 1997a, 3-7). Upstream inputs, especially of PCBs, make some control efforts more difficult than others, and the existence of a successful River Rouge RAP that discharges into the Detroit River has taken attention away from their international connecting channel. The magnitude of the pollution problem is shown in this statistic: there are 14,300 commercial and industrial dischargers connected to the Detroit and Windsor sewage treatment plants, 13,000 of which are on the Detroit side alone.

The institutional design was the responsibility of the Michigan Department of Natural Resources (MDNA) under an agreement with the EPA and the Ontario and Canadian governments. This department, along with six other agencies, began the task of writing Stage 1 in 1986 and chose to add four members of a binational public advisory council (BPAC) to provide it with advice. The BPAC consisted of 40 stakeholders – 20 from each side of the border – chosen by the MDNR. (The original Windsor participants were suggested by consultants for the Ontario Ministry of the Environment; later participants were elected by BPAC members.) The Stage 2 RAP was developed through four technical work groups that contained both agency and BPAC members. The stakeholders involved in these processes included few property owners or representatives of major sectors that could constitute partners for remediation or even clients for regulation. The efforts of BPAC members to expand their roles, for example by prioritizing remediation strategies, were vetoed by the MDNR on the grounds that "the role of the BPAC is advising as stated in its charge" (Susan Bouzie, MDNR, Minutes of the 29th Meeting of the Detroit River Public Advisory Council, Windsor, 26 August 1992, 7; quoted in Strutt 1993, 27). In 1997, the Michigan department withdrew from its lead agency role and identified the BPAC as the primary implementer, notwithstanding its flawed composition (IJC 1997a, 8). At that time, the remediation of the AOC rested on the environmental efforts of stakeholders acting independently. At the active intervention of the IJC, the four major governments for the connecting channel recommitted themselves to a RAP in April 1998. However, the Michigan Department of Environmental Quality (now the Michigan lead agency) will focus on developing implementation agreements that will empower local implementers to move the process forward. However, specific local recommendations and implementation actions from the American side are still awaited.

In contrast, on the Canadian (Windsor) side, a newly constituted multistakeholder group called the Detroit River Cleanup Committee was formed in 1998 to coordinate remediation. This signalled a move away from a coordinated binational approach to a "two-sides-of-the-river-policy" like the Niagara River RAP. An extensive inventory of completed and required recommended actions now exists (Environment Canada 2000).

The Toronto experience is roughly similar, albeit without the added complexity of international fragmentation. The Toronto AOC is essentially an administrative rather than a geographical unit as it encompasses six watersheds and a population exceeding 4 million. As in the Detroit River case, the Toronto Stage 1 report (Environmental Conditions and Problem Definition) was drafted in 1988 by an intergovernmental writing team consisting of agency personnel (prior to the formation of a public advisory committee). The Ontario and Canadian environmental agencies, charged with institutional design, thus saw themselves as the control centre of this AOC. The Ontario Ministry of the Environment formed ten public advisory committees, based on sectors, and these committees selected members for an area-wide public advisory committee. At the same time, the federal government went into competition with its own RAP by establishing the Royal Commission on the Future of the Toronto Waterfront, which later became the permanent Waterfront Regeneration Trust. It duplicated the planning activities of the RAP, but with significantly more resources and publicity. The sectoral committees gradually disintegrated, and while a draft Stage 2 report was produced (on implementation strategies), the Ontario government vetoed both its implementation and its coordination. Between 1994 and 1997, the Toronto RAP consisted largely of the uncoordinated individual efforts of some of the stakeholder organizations.

Because of pressure from both the public and Environment Canada, a new effort to revive the Toronto RAP began in 1997. This time the Waterfront Regeneration Trust and the Toronto and Region Conservation Authority (a special regional board charged primarily with flood control management) were designated as lead agencies for Environment Canada and the Ontario Ministry of the Environment. Instead of organizing stakeholders by sector, which only intensified resource-use conflicts between sectors, the stakeholders are now organized into watershed and waterfront organizations. These include the Toronto Bay Initiative, the Humber Alliance, the Don Watershed Regeneration Council, the Rouge Park Alliance, and various subwatershed (creek) associations. New recommended actions – an implementation strategy that parties would accept – based on these watersheds and waterfronts have now been adopted, and special federal appropriations associated with the unsuccessful bid for the 2008 Olympics have been dedicated to the Don Valley watershed. This experience suggests that the institutional designers may have learned from experience about some of the institutional obstacles associated with large-scale multiple watershed sites, not to mention about the limitations of top-down controls. As late as 1994, one PAC member was concluding that "the bureaucracy by and large uses consultant processes as a means of adding or subtracting credibility from the issues as they choose" (cited in Gunther-Zimmerman 1994, 229).

Two large-scale RAPs, already referred to as successes, belie the proposition that size (and thus complexity) make institutional design inoperable. These are RAPs for Hamilton Harbour on Lake Ontario and for the Cuyohoga River, which flows into Lake Erie at Cleveland, Ohio.

In the Hamilton case, we have already noted how a group of 50 stakeholders formulated the Stage 1 and Stage 2 reports using the government agency writing team as technical support. By way of institutional design, it set up two bodies to oversee implementation, which was to be conducted by 14 agencies and organizations, mostly in pooled coupling strategies. The implementers formed the Bay Area Implementation Team, which was serviced by a RAP coordinator who was also an employee of Environment Canada. The other stakeholders were grouped into a Bay Area Restoration Council that would produce annual audits of progress towards implementation and that would run education awareness programs and related activities. The council has expanded into 160 paid members, both individual and organizational, and receives funding from a variety of sources. Its power over the implementation process is solely that of information and prestige. The Ontario Ministry of the Environment vetoed an original scheme to make the RAP a legal document and, hence, make implementers subject to injunctions on the part of other stakeholders. In sum, the stakeholders were and are successfully organized despite the difficulties associated with a watershed population of 500,000 and ten impaired beneficial uses.

The Cuyahoga River RAP parallels the Hamilton Harbour case. The Ohio Environmental Protection Agency designed a coordinating committee consisting of 35 stakeholders and gave it express authority to develop a Stage 1 plan. Stage 1 was completed in 1992, and a draft Stage 2 is now being completed. The AOC covers the lower 45 miles of the river plus ten miles of the Lake Erie shoreline. The river drains an agricultural and urban watershed of 1.6 million people, including the industrial cities of Cleveland and Akron, and it achieved notoriety in 1969 when it caught on fire. It has some ten impaired uses (like Hamilton) so its ecosystem difficulties are many, ranging from conventional problems such as low dissolved oxygen levels in the water to contaminated sediments from urban runoff (especially PCBs) and rural non-point sources (especially pesticides). The implementation report, Stage 2, again involves stakeholder committees working with technical support staff from government agencies. The Cuyahoga River Community Planning Organization, funded by three local private foundations, develops programs for public involvement, education, and research and is roughly comparable to the Hamilton Bay Area Restoration Council.

In this case, size – as indicated by population, industrialization, and impaired uses – is accommodated through careful institutional design. While size often indicates complex environmental problems, it need not present insuperable obstacles to remediation. Deconcentrated planning and well-crafted and adaptable implementation strategies appear to be necessary ingredients for successful RAPs, including the larger AOCs.

Conclusions

The prevailing patterns of behaviour found in this "Great Lakes experiment," which mobilized stakeholders to adjust their property rights to restore beneficial uses, are disappointing. State and provincial environment agencies, in collaboration with their federal counterparts, were empowered to design institutional arrangements for RAPs. They used this opportunity to construct stakeholder organizations to advise them how to plan and implement remedial improvements. They then simply layered RAPs onto their pre-existing programs of environmental management and pollution control. Consequently, little has changed because of RAPs, other than restoration efforts being given legitimacy and/ or involved stakeholders being disappointed with the processes. In many cases, it appears as if the environmental agencies in question were/are indifferent to, or cavalier about, institutional design rather than being openly hostile. But the prevailing patterns evolved nevertheless.

There are important exceptions to these prevailing patterns. We have noted the successes of Hamilton Harbour and the Cuyahoga River in organizing stakeholders in relatively large and polluted AOCs. In these cases, and there could be others, stakeholders set goals and made trade-offs among their interests in order to facilitate some kind of efficient solution to multiple-use conflicts. In these cases, the relevant government agencies engaged in institution building and design in order to get stakeholders to work together. The consensual arrangements provided productive solutions at the planning level.

Consensual arrangements can work well to ensure that all representative interests can be aggregated into a commonly planned decision or decisions. They work less well at the implementation stage as they create incentives for opportunism. We saw this in the Randles Reef issue of Hamilton Harbour. It was also exhibited in the Port Hope RAP, where the Atomic Energy Commission of Canada followed a non-cooperative strategy and exercised its monopoly power as an atomic regulatory agency.

We found that consensual arrangements were more difficult to establish in AOCs that spanned international and interstate boundaries. In the cases we examined – the Niagara River, the St. Lawrence River, and the Menominee River – parallel RAPs had to be developed to deal with differences in priorities from one side of the river to the other side. In the Ontario and New York RAPs, these parallel arrangements were in the form of separate official plans. In the Menominee case, Wisconsin and Michigan had to divide planning recommendations and implementation strategies. And, it also appears the same process is evolving on the Detroit River, where the indifference of the Michigan government (again) contrasts with renewed efforts on the Windsor, Ontario, shorelines and creeks. Currently, the Detroit River appears to have a minimal chance of remediation.

There seems to be some evidence that the institutional designers can and are learning from experience. The totally revamped Toronto RAP is a case in point, as it replaces sectoral committee structures and recommendations with those based on watersheds and waterfronts. The IJC appears anxious to promote success stories from selected RAPs.

In general, we can state that RAPs have done little to alter the behaviour of stakeholders and their bundles of property rights. Institutions, or structures, or rules (whatever term one wishes to use) are, essentially, about incentives. They create incentive systems that render certain behaviours costly and others less so. RAPs were constructed without sensitivity to incentive systems and their relevance to environmental management.

7
Conclusion:
Promises and Performances

In the 1970s, governments and industries around Lake Erie began the costly but successful process of reversing the eutrophication of this, the shallowest of the Great Lakes. Concerted and large-scale actions were undertaken in a resource shared by two countries, four states, and one province. A large-scale common-pool problem was capable of resolution across national and subnational jurisdictions. The International Joint Commission was both a catalyst and a facilitator of this concerted effort.

At the same time, environmental scientists and regulators were discovering some more intractable problems around the Great Lakes. Rivers, bays, and harbours were suffering from environmental degradation because of different kinds of pollution and diverse kinds of resources exploitation. Many uses of these sites had been, and were being, curtailed or eliminated. The flames that burned in 1978 on the Cuyohoga River in Cleveland, Ohio, symbolized these new priorities.

The IJC's Water Quality Board began to assemble information about resource degradation in sites around the Great Lakes basin. As early as 1974, the Water Quality Board identified 69 "problem areas" as being in need of remediation (Colborn et al. 1990, 201). By 1983, it had identified 43 "Areas of Concern" and established common criteria across all of these areas to indicate the scope of their degradation. The criteria were referred to as "impaired beneficial uses," and, in Chapter 1, we have noted how the IJC acknowledges the different impairments in each AOC.

Progress in restoring the beneficial uses is, after 15 years, mixed. Some AOCs have made substantial progress, and one has been delisted. Yet there are others where remediation has not begun, and still others where efforts are partial, sporadic, and incomplete. We need to revisit the promises and practices of remediation before we address the performances that comprise these mixed conclusions.

The Promises

The RAP program was launched with immense hyperbole from officials and from academic scientists. In 1990, William K. Reilly spoke of the RAP process as

> an example for the United States, Canada and the world. It provides planning that is geographically focused, ecosystem based, and cuts across environmental media. [It] offers a model of how to make better use of existing programs while determining the other actions needed to restore human and ecosystem health. By including stakeholders, it builds support and creates accountability. It provides a model for successful problem solving and a practical way to attain the goals of the Great Lakes Water Quality Agreement to restore and maintain the chemical, physical and biological integrity of the waters of the Great Lakes Basin Ecosystem (Hartig and Zarull 1992, 4).

As late as 1995, public officials, striving to publicize "the good news" about RAPs, could write:

> Incorporating an ecosystem approach into RAPs has meant viewing different organizations, agencies and stakeholders as equal members of a team in a partnership to identify and solve environmental problems ... Sharing decision-making power and accepting responsibility for action is requisite, as no single agency or organization has the capacity to plan and implement RAPs ... Securing broad based participation in the RAP process takes advantage of the knowledge and expertise that community leaders possess about their Area of Concern and helps create the necessary sense of ownership to establish the political will for plan implementation (Hartig et al. 1995, 8).

The irony is that these aspirations do indeed roughly match the operations of the successful RAPs examined in this study. But they do not spell out the necessary conditions for designing the processes needed to realize the aspirations. The necessary conditions are essentially those of "incentive systems." People respond to incentives, positive and negative, that are structured by rules and learned through experience and peer example. RAP coordinators, for example, who are either employed by a lead environmental agency or hired under contract, have one obvious set of incentives: to respond to the "commands" or suggestions of their managing supervisors. They also have incentives to respond to the RAP participants who may be drawn from a variety of stakeholder organizations. They must thus manage horizontally as well as vertically. In these circumstances, it would be tempting to adopt institutional arrangements that would ensure that only a limited number of participants – who could agree upon goals, recommendations, and implementing strategies of action – would be deemed relevant. Thus, the managers (vertical) of the RAP coordinators would be satisfied and their stakeholders could operate, with limited transaction costs, in a horizontal management fashion. In this way, the inclusiveness of the stakeholder process (as one institutional concern) can be compromised.

The promises, however, should not be underestimated. The Great Lakes have been a catalyst for economic development, industry, and recreation for almost two centuries. We also saw, in Chapter 2, that, given diminishing shipping and fishing uses as well as the early exploitation of most hydroelectric power sites, the primary use of the Great Lakes nowadays is for waste disposal. Within this context, efforts to fashion and implement plans to restore beneficial uses in the worst polluted areas were both necessary and promising.

There was further optimism because environmental groups, industry, municipalities, and other actors who were not regularly part of the "water quality management agreement process" were now going to be included in plan formulation. The IJC recommended to the two national governments that the RAP processes include stakeholders, in part because it believed that the stakeholders could act as "scrutineers" of the RAP process. It would therefore be more difficult for governments, especially state and provincial governments, to escape their responsibilities (IJC Commissioner, personal interview, October 1999).

The IJC in fact only specified what we might call "implicit constitutional rules" to frame the collective-choice processes for the 43 RAPs. One of these rules was that the respective governments would agree to include multiple stakeholders in the RAP processes. The second was that the respective federal, state, and provincial governments could designate one or two lead agencies to fashion the institutional design. (This rule was implicit in that the "normal" government processes involving the IJC included [1] recommendations of the IJC to the two national governments and [2] agreements between the national governments and their provincial/state member governments to implement the recommendations.) The third implicit rule was that the lead agencies were constrained by the understanding that they could not unilaterally design RAPs or develop implementation strategies. The lead agencies were, however, granted powers to choose stakeholders and to design the rules under which their interests would be included.

The promises of RAPs were thus extensive. New efforts were to be made to deal with pollution, especially of the complex kinds that would be stressful to the ecosystems. These efforts were to be focused on key AOCs that manifested the major problems, posed in terms of impaired beneficial uses. Governments would enlist the help of environmentalists and industrialists and other community members to ensure that the communities in the AOCs would be responsible for remediation. And the stakeholders, in turn, would be empowered to develop their own collective-choice processes to target the stresses of their ecosystems.

Performance

We examined the results of RAPs in Chapter 6. In general, they have had modest success: some RAPs have performed well, with demonstrable effects on restoring beneficial uses; others have not performed well, due to institutional and/or ecosystem difficulties.

We may put forth some of the propositions we discovered through our investigations:

(1) the wider the scope of stakeholder representation, the stronger the performance of the RAP
(2) the more consensual the plan formulation process, the better the RAP performance
(3) however, the greater the number of sovereign jurisdictions involved in plan formulation, the more difficult it is to construct and implement a RAP, and the greater the probability of poor performance
(4) the fewer the interdependencies in plan implementation (as opposed to formulation), the greater the performance
(5) sequential and reciprocal interdependencies in plan implementation were more likely to produce poor performance than were pooled interdependencies
(6) the greater the complexity and number of ecosystem uses (frequently associated with size), the poorer the performance.

These findings, however, are contingent upon the lead agencies fashioning an appropriate institutional design. Lead agencies were engaged in a collective-choice process with little understanding of the wider significance of institutional designs for successful RAP performance. "Getting the science right," which is often a watchword of lead environmental agencies, was not partnered with, for example, "getting the social science right"; instead, many agencies adopted a working assumption that held that the most successful RAPs would be those in which the stakeholders would help the lead agencies plan, finance, and implement their legislative authority for the AOCs. In other words, bureaucratic self-interest dominated the collective-choice process. Since the collective-choice process was so narrow, it was the interests of the lead agency (or agencies) that prevailed.

Yet there were some exceptions to this constitutional design pattern. Sometimes the monopoly or quasi-monopoly position of the lead agencies gave them sufficient discretion to choose other options. In one case, that of Hamilton, the institutional design was an accident. The lead agencies had no models to copy and hired a consultant who designed the arrangements to include all relevant stakeholders, with an open agenda for decision making. The province never permitted this design to proceed further.

We can thus attribute part of the success, or part of the failure, of the Great Lakes' RAPs to institutional design issues. From a practical standpoint, it is hoped that the IJC and other bodies will learn from the RAP experience and think through the incentive systems for different constitutional and collective-choice processes. Bureaucratic lead agencies currently have neither the interest nor the capacity to engage in the design of well functioning collective-choice processes. If this system of governance persists, then the lakes will not be restored.

We are now in a position to reassess our theory and explanations about common-property resources in order to take account of our findings concerning this multiple-use and large-scale ecosystem.

Theoretical Review

In Chapter 4 we presented the theory of common-property resources in order to better understand the logic and dynamics of user group and stakeholder interdependencies. We explored how RAPs could pick up these interdependencies and fashion "solutions" to the remediation of CPRs in AOCs. We noted three important concerns pertaining to large-scale multiple-use CPRs such as the Great Lakes:

(1) There are multiple and dynamic interdependencies inherent in ecosystems and, hence, comparable interdependencies for stakeholders that enjoy or use these ecosystems.
(2) Rules devised to allocate and distribute bundles of rights among stakeholders are subject to incentives to change because of these multiple and dynamic ecosystem interdependencies.
(3) Collective-choice mechanisms could, potentially, fashion newer operational rules to take account of these incentives. The newer

rules could build upon common and statutory law methods of adaptation (cases, courts, precedents) that have demonstrated some limitations in balancing different uses and in sustaining ecosystems from place to place.

We can now add a fourth major concern pertaining to large-scale multiple-use common-pools situations:

(4) Bureaucracies are inappropriate designers of collective-choice mechanisms in that their incentives may or may not coincide with those needed to balance multiple users and to sustain ecosystems from situation to situation.

All four concerns form the context from which CPR theory may be extended. Three features of this extension are suggested below:

(1) those concerning property rights "bundles" in CPR situations
(2) those concerning the design of collective-choice mechanisms for interdependent users
(3) those concerning the incentives on designers of collective-choice arrangements

Property Rights and Stakeholders
We noted previously that stakeholders in RAPs could be, and were in practice, individuals, clubs, corporations (private, public, or government), and government departments (national, regional, or local). They reflected wider societal interests that had a stake in the outcomes made possible by RAPs. In most cases, they were direct users of the resource – for waste disposal, container shipping, recreation, irrigation, fishing, and so on. They could also be environmentalists, who place intrinsic rather than instrumental value on the common property.

Using the typology suggested by Schlager and Ostrom (1992), the stakeholders could be owners, proprietors, claimants, authorized users, and (suggested by us) squatters.

The legal scholars Hohfeld (1919) and Commons (1924) clarified the nature of property rights (such as management) by demonstrating that they are not rights to things, as such, but concern the claims and duties that persons have with respect to other persons, "even if these claims and duties are deployed over the uses and dispositions of things" (Ali 2000, 5).

Hohfeld is significant to our analysis of common property in the Great Lakes in that he points out how many of the users (and stakeholders) have privileges as authorized users but do not have reciprocal duties. There may be conditions attached to an authorization to use the lakes, bays, and rivers for the disposal of liquid industrial effluent, but there may be no reciprocal duty to or from other users. In contrast, riparians may have claims on waste dischargers, but these can be attenuated by the conditions of use that may be authorized for in a discharge. The Ontario Environmental Protection Act, for example, removes the common law damage claims that riparians may have with respect to pollution caused by municipal waste water treatment plants. We can have many thousands of users of the Great Lakes for one or more of the uses we have depicted, but many of the users are not made legally interdependent (i.e., they are not made claim holders) with regard to the "technical" interdependencies they may create.

Let us work through this again using the case of a recreational fisher. This person typically requires a licence from a state or provincial agency to gain access to the waters of the Great Lakes and to withdraw fish. There may be conditions attached to the licence, like a quota on the number of particular species that may be harvested in one day. Unless the fisher is a riparian, he or she has no claim on others who fish, discharge wastes, watch birds, or own freighters. The fisher simply has an authorization to use the waters for the purpose of fishing; but neither the fisher nor others have claims with regard to the fishing.

In cases like these, it is the responsibility of the government agencies that issue the licence to police for violations. Uses are not self-formed but, rather, are governed by those with the power to create and remove claims. Obviously, policing is more difficult across sovereign jurisdictions like nation states or states and provinces.

If this analysis is correct, then authorized users (or squatters) have privileges that do not require them to take other users into account (unless so specified in the conditions of authorized use). They can, of course, voluntarily agree to a correlative duty. In these circumstance, it will depend on the values of the person and on the social norms of the community whether voluntary self-regulation occurs. In large group situations, we know that social norms tend to dissipate and that free riding occurs (Olson 1965).

In the large multiple-use legal system that has evolved on the Great Lakes, reciprocal duties on users are either not specified in their conditions of use or they are specified and then policed (somewhat) by government. In the former case, voluntary reciprocity can potentially take place. However, the Great Lakes contain 32 million people and countless thousands of users. With these conditions, the Great Lakes resemble a large, partially regulated commons whose negative interdependencies have come to dominate the ecosystems. In these circumstances, we can expect failure in managing and balancing multiple uses. Users have little incentive to take other users into account. Degraded bays, harbours, and river mouths are predictable results. However, there are collective-choice mechanisms that may change these incentives and consequences.

Collective Choices

It is through collective choices that the privileges and claims of users of the Great Lakes may be altered. To use Hohfeld's terminology, collective-choice forums can change the powers and immunities of users.

RAPs had the potential to change powers and immunities. However, they were designed to maintain the pre-existing powers and immunities, and to see if the voluntary and largely consensual coordination of users and managers could move the stakeholders towards better performance and, hence, restored beneficial uses. They were and are an exercise in horizontal management and in voluntary reciprocity between stakeholders (Sproule-Jones 2000). No attempts were made by the respective governments to alter the legal arrangements through which RAPs would be fashioned and implemented. (The US Critical Programs Act, 1990, which authorized RAPs, did not alter the pre-existing regulatory regime of American laws.)

Thus, RAPs were modest experiments in collective choice. National, state, and provincial governments were unwilling to fundamentally alter the powers and immunities of the stakeholders. The mixed performance of RAPs that was traced in the previous chapter is partly explained by these collective-choice limitations.

Nevertheless, RAPs performed differently and with different degrees of effectiveness depending upon their designs for formulation and implementation. They were constructed by the lead environmental agencies in each state and, jointly, by the Canadian federal and provincial governments in the Ontario cases. The model that prevailed in many RAPs was the one that included those users whom the agencies perceived to be important advisors as they went about analyzing their perceived site-specific problems and coordinating their different solutions. This was not a model that would produce the best feasible collective choices within the context of stakeholders' pre-existing bundles of property rights. Occasionally, however, the model produced an inclusive group of stakeholders and (financial and legal) opportunities to implement the plan on the part of individual stakeholders or groups of stakeholders.

Further, some RAPs were able to build upon voluntary goodwill among participating stakeholders and use the social capital to coordinate and implement new programs. Hamilton and Cleveland come to mind. However, with regard to communities that had little experience in collective stakeholder activities, voluntary cooperation was insufficient. This was particularly true in international waterways like the St. Lawrence, Niagara, or Detroit Rivers, whose cross-national interactions have been infrequent and not particularly productive. Further, AOCs were identified with no social criteria in mind. They were, in fact, largely an accident of water sampling activity – which is hardly an ecosystem criterion. No due care and attention was given to these boundary issues, human or otherwise. Like so much of the RAP story, governments were cavalier with regard to institutional design issues.

The key to understanding large-scale multiple-use common properties, their successes and failures, seems to lie with collective-choice processes that can alter legal relationships among the various users of the ecosystems in question. Bureaucracies, given their own agendas and mandates, are inappropriate designers and should not be charged with these tasks. It would seem that some forms of self-governing collective-choice processes would be more appropriate (V. Ostrom 1997).

Designs and Designers
It would appear, in many situations, that government agencies had neither the skill nor the interest to construct collective-choice mechanisms for successful RAP formations and implementation. This is ironic as many agencies saw themselves as mandated to remediate AOCs (albeit with stakeholder advice). Stakeholders would be advisors and interest groups rather than co-managers of AOCs. With this perspective, the transaction costs of plan formation and implementation could be minimized. Effective solutions were anticipated.

In multiple-use resource situations, however, a government agency is rarely "in charge" and rarely has sufficient expertise and powers to construct and implement effective resource plans. In other words, property rights are scattered and variable, and the comparative advantage of government agencies appears to lie in their powers to facilitate, coordinate, and partially finance particular resource problems among users. This comparative advantage can be layered onto their regulatory powers over some resource uses. They are important but not omnipotent stakeholders. They are also accountable to governments whose territories exceed any one AOC but that are not extensive enough to contain all AOCs. Agency agendas are set, at best, with other concerns competing prominently with those focused upon remediating impaired beneficial uses in different locations.

However, governments have a body of accumulated knowledge about institutional designs for collective choices. This knowledge could be drawn upon in a way similar to drawing upon environmental science for resource decision making. The institutional design knowledge is arguably comparable in predictive capacity to that of environmental science. It is ignored, in part, because decision makers believe that social science knowledge can be grasped intuitively, without scholarly examination. It is more usually ignored because decision makers presume that they are in charge of, rather than co-managers of, a common property.

The body of accumulated knowledge about institutional designs began to be amassed in the eleventh and twelfth centuries when the first written constitutions were established for European cities, especially in The Netherlands and Italy (Berman 1983). Those constitutions spelled out, among other things, the participatory relationships among members of the community. These municipal constitutional laws were one of a number of codified legal systems (e.g., common law, canon law, mercantile law, and royal law). There were occasional disputes among these rival legal arrangements, and, in some countries (e.g., England), royal law became supreme over the course of four centuries. Royal law was later transformed into statutory law, as parliaments replaced the Crown as a source of law. Canada inherited a tradition of royal law, and this still dominates ideas about the place of parliaments (provincial and federal). The United States also incorporated some major elements of royal law into its democratized state and federal governments (Sproule-Jones 1984). The municipal tradition of self-governance was an acknowledged part of American governance arrangements in the nineteenth century, as Toqueville discovered. But it, too, has been ameliorated over the last century (V. Ostrom 1997).

With this theoretical perspective and its extensions, the designs for RAPs become more understandable, varying, as they do, between (1) stakeholder empowerment in the formation and implementation of plans and (2) limited stakeholder advice and involvement for the lead agencies of American state and Canadian provincial and federal governments. The designs can best be viewed as experiments, and this book can best be viewed as an effort to understand and learn from these experiments. We hope governments can also learn from them.

Conclusion

The promises of RAPs and remediation on the Great Lakes were many and important. The performances were modest; however, the potential remains great. Governments in both countries need to set rules that collective-choice bodies can use to change the legal relationships of users and stakeholders on the Great Lakes. Adaptation is necessary because the dynamics of the ecosystems under use are adapting and changing. The story of the Great Lakes and its governance is far from over. We still await their restoration. And we must hope that the fragile ecosystems can wait too, until human beings have fashioned an acceptable way of integrating themselves into the environment.

Appendix A

Guidelines for recommending the listing and delisting of Great Lakes Areas of Concern

Use impairment	Listing guideline	Delisting guideline	Rationale	Reference
Restrictions on fish and wildlife consumption	When contaminant levels in fish or wildlife populations exceed current standards, objectives, or guidelines, or public health advisories are in effect for human consumption of fish or wildlife. Contaminant levels in fish and wildlife must be due to contaminant input from the watershed.	When contaminant levels in fish and wildlife populations do not exceed current standards, objectives, or guidelines, and no public health advisories are in effect for human consumption of fish or wildlife. Contaminant levels in fish and wildlife must be due to contaminant input from the watershed.	Accounts for jurisdictional and federal standards; emphasizes local watershed sources.	Adapted from Mack (1988).
Tainting of fish and wildlife flavour	When ambient water quality standards, objectives, or guidelines for the anthropogenic substance(s) known to cause tainting are being exceeded or survey results have identified tainting of fish or wildlife flavour.	When survey results confirm no tainting of fish or wildlife flavour.	Sensitive to ambient water quality standards for tainting substances; emphasizes survey results.	See American Public Health Association (1980) for survey methods.

▼ *Appendix A*

Use impairment	Listing guideline	Delisting guideline	Rationale	Reference
Degraded fish and wildlife populations	When fish and wildlife management programs have identified degraded fish or wildlife populations due to a cause within the watershed. In addition, this use will be considered impaired when relevant, field-validated, fish or wildlife bioassays with appropriate quality assurance/quality controls confirm significant toxicity from water column or sediment contaminants.	When environmental conditions support healthy, self-sustaining communities of desired fish and wildlife at predetermined levels of abundance that would be expected from the amount and quality of suitable physical, chemical, and biological habitat present. An effort must be made to ensure that fish and wildlife objectives for Areas of Concern are consistent with Great Lakes ecosystem objectives and Great Lakes Fishery Commission fish community goals. Further, in the absence of community structure data, this use will be considered restored when fish and wildlife bioassays confirm no significant toxicity from water column or sediment contaminants.	Emphasizes fish and wildlife management program goals; consistent with Agreement and Great Lakes Fishery Commission goals; accounts for toxicity bioassays.	Adapted form Manny and Pacific (1988); Wisconsin DNR (1987); United States and Canada (1987); Great Lakes Fishery Commission (1980).
Fish tumours or other deformities	When the incidence rates of fish tumours or other deformities exceed rates at unimpacted	When the incidence rates of fish tumours or other deformities do not exceed rates at unimpacted	Consistent with expert opinion on tumours; acknowl-	Adapted from Mac and Smith (1988); Black (1983);

Use	Impaired when	Restored when		Reference
	control sites or when survey data confirm the presence of neoplastic or preneoplastic liver tumours in bullheads or suckers.		edges background incidence rates.	Baumann et al. (1982).
Bird or animal deformities or reproductive problems	When wildlife survey data confirm the presence of deformities (e.g., cross-bill syndrome) or other reproductive problems (e.g., egg-shell thinning) in sentinel wildlife species.	When the incidence rates of deformities (e.g., cross-bill syndrome) or reproductive problems (e.g., egg-shell thinning) in sentinel wildlife species do not exceed background levels in inland control populations.	Emphasizes confirmation through survey data; makes necessary control comparisons.	Adapted from Kubiak (1988); Miller (1988); Wiemeyer et al. (1984).
Degradation of benthos	When the benthic macro-invertebrate community structure significantly diverges form unimpacted control sites of comparable physical and chemical characteristics. In addition, this use will be considered impaired when toxicity (as defined by relevant, field-validated bioassays with appropriate quality assurance/quality controls) of sediment-associated contaminants at a site is significantly higher than controls.	When the benthic macro-invertebrate community structure does not significantly diverge from unimpacted control sites of comparable physical and chemical characteristics. Further, in the absence of community structure data, this use will be considered restored when toxicity of sediment-associated contaminants is not significantly higher than controls.	Accounts for community structure and composition; recognizes sediment toxicity; uses appropriate control sites.	Adapted from Reynoldson (1988); Henry (1988); IJC (1988).

▼ *Appendix A*

Use impairment	Listing guideline	Delisting guideline	Rationale	Reference
Restrictions on dredging activities	When contaminants in sediments exceed standards, criteria, or guidelines such that there are restrictions on dredging or disposal activities.	When contaminants in sediments do not exceed standards, criteria, or guidelines such that there are restrictions on dredging or disposal activities.	Accounts for jurisdictional and federal standards; emphasizes dredging and disposal activities.	Adapted from IJC (1988).
Eutrophication or undesirable algae	When there are persistent water quality problems (e.g., dissolved oxygen depletion of bottom waters, nuisance algal blooms or accumulation, decreased water clarity, etc.) attributed to cultural eutrophication.	When there are no persistent water quality problems (e.g., dissolved oxygen depletion of bottom waters, nuisance algal blooms or accumulation, decreased water clarity, etc.) attributed to cultural eutrophication.	Consistent with Annex 3 of the Agreement; accounts for persistence of problems.	United States and Canada (1987).
Restrictions on drinking water consumption or taste and odour problems	When treated drinking water supplies are impacted to the extent that (1) densities or disease-causing organisms or concentrations of hazardous or radioactive toxic chemicals or radioactive substances exceed human health standards, objectives, or guidelines, (2) taste and odour	For treated drinking water supplies: (1) when densities of disease-causing organisms or concentrations of hazardous or toxic chemicals or radioactive substances do not exceed human health objectives, standards, or guidelines; (2) when taste and odour problems are absent; and	Consistency with the Agreement; accounts for jurisdictional standards; practical; sensitive to increased cost as a measure of impairment.	Adapted from United States and Canada (1987).

	problems are present; or (3) treatment needed to make raw water suitable for drinking is beyond the standard treatment used in comparable portions of the Great Lakes, which are not degraded (i.e., settling, coagulation, disinfection).	(3) when treatment needed to make raw water suitable for drinking does not exceed the standard treatment use in comparable portions of the Great Lakes that are not degraded (i.e., setting, coagulation, disinfection).		Adapted from United States and Canada (1987); Ontario Ministry of the Environment (1984).
Beach closings	When waters that are commonly used for total-body contact or partial-body contact recreation exceed standards, objectives, or guidelines for such use.	When waters that are commonly used for total-body contact or partial-body contact recreation do not exceed standards, objectives, or guidelines for such use.	Accounts for use of waters; sensitive to jurisdictional standards; addresses water contact recreation; consistent with the Agreement.	
Degradation of aesthetics	When any substance in water produces a persistent objectionable deposit, unnatural color or turbidity, or unnatural odour (e.g., oil slick, surface scum).	When the waters are devoid of any substance that produces a persistent objectionable deposit, unnatural color or turbidity, or unnatural odour (e.g., oil slick, surface scum).	Emphasizes aesthetics in water; accounts for persistence.	Adapted from the Ontario Ministry of the Environment (1984).

▼ *Appendix A*

Use impairment	Listing guideline	Delisting guideline	Rationale	Reference
Added costs to agriculture or industry	When there are additional costs required to treat the water prior to use for agricultural purposes (i.e., including, but not limited to, livestock watering, irrigation, and crop-spraying) or industrial purposes (i.e., intended for commercial or industrial applications and noncontact food processing).	When there are no additional costs required to treat the water prior to use for agricultural purposes (i.e., including, but not limited to, livestock watering, irrigation, and crop-spraying) and industrial purposes (i.e., intended for commercial or industrial applications and noncontact food processing).	Sensitive to increased cost and a measure of impairment.	Adapted form the Michigan DNR (1977).
Degradation of phytoplankton and zooplankton populations	When phytoplankton or zooplankton community structure significantly diverges from unimpacted control sites of comparable physical and chemical characteristics. In addition, this use will be considered impaired when relevant, field-validated phytoplankton or zooplankton bioassays (e.g., *Ceriodaphnia*, algal fractionation bioassays) with appropriate quality assurance/quality controls confirm toxicity in ambient waters.	When phytoplankton and zooplankton community structure does not significantly diverge from unimpacted control sites of comparable physical and chemical character-istics. Further, in the absence of community structure data, this use will be considered restored when phytoplankton and zooplankton bioassays confirm no significant toxicity in ambient waters.	Accounts for community structure and composition; recognizes water column toxicity; uses appropriate control sites.	Adapted form IJC (1987).

| Loss of fish and wildlife habitat | When fish and wildlife management goals have not been met as a result of loss of fish and wildlife habitat due to a perturbation in the physical, chemical, or biological integrity of the Boundary Waters, including wetlands. | When the amount and quality of physical, chemical, and biological habitat required to meet fish and wildlife management goals have been achieved and protected. | Emphasizes fish and wildlife management program goals; emphasizes water component of Boundary Waters. | Adapted from Manny and Pacific (1988). |

Source: Modified from International Joint Commission (1991b, 4-5). For references cited above, refer to the original document.

Appendix B

Listing and delisting guidelines for Great Lakes Areas of Concern with examples of quantitative objectives and targets for use restoration

Use impairment	Listing and delisting guidelines	Example of quantitative objective/target for use restoration
Restrictions on fish and wildlife consumption	*Listing Guideline:* When contaminant levels in fish or wildlife populations exceed current standards, objectives, or guidelines, or public health advisories are in effect for human consumption of fish or wildlife. *Delisting Guideline:* When contaminant levels in fish and wildlife populations do not exceed current standards, objectives, or guidelines, and no public health advisories are in effect for human consumption of fish or wildlife. (*Note:* Contaminant levels in fish and wildlife must be due to contaminant input from watershed.)	Over 159,000 kg of PCBs reside in Kalamazoo River (Michigan) sediments and have resulted in contamination of the fishery. Two levels of cleanup standards apply: • a short-term target based on the U.S. Food and Drug Administration Action Level of 2 mg/kg PCBs in the edible portion of fish • a long-term target of 0.05 mg/kg PCBs in fish tissue established to protect human health through Rule 57 of Michigan Water Quality Standards (Waggoner and Creal 1992).
Tainting of fish and wildlife flavour	*Listing Guideline:* When ambient water quality standards, objectives, or guidelines for the anthropogenic substance(s) known to cause tainting are being exceeded or survey results have identified tainting of fish or wildlife flavour. *Delisting Guideline:* When survey results confirm no tainting of fish or wildlife flavour.	In Spanish River (Ontario), 72 hour *in situ* fish exposure under low flow and subsequent sensory evaluation were used to re-evaluate fish tainting due to mill effluent (upstream control site and downstream effluent plume). A triangle test (three samples to each of eleven panelists; two samples the same and one different) was used to determine a difference (Jardine and Bowman 1992). The number of correct responses must not be significantly different (95% confidence) from chance of guessing odd sample. Based on this approach, a sensory panel could not distinguish tainting in fish exposed to mill effluent.

Degraded fish and wildlife populations

Listing Guideline: When fish and wildlife management programs have identified degraded fish or wildlife populations due to a cause within the watershed. In addition, this use will be considered impaired when relevant, field-validated fish or wildlife bioassays with appropriate quality assurance/quality controls confirm significant toxicity from water column or sediment contaminants.

Delisting Guideline: When environmental conditions support healthy, self-sustaining communities of desired fish and wildlife at predetermined levels of abundance that would be expected from the amount and quality of suitable physical, chemical, and biological habitat present. An effort must be made to ensure that fish and wildlife objectives for Areas of Concern are consistent with Great Lakes ecosystem objectives and Great Lakes Fishery Commission fish community goals. Further, in the absence of community structure data, this use will be considered restored when fish and wildlife bioassays confirm no significant toxicity from water column or sediment contaminants.

In Hamilton Harbour (Lake Ontario), the overall objective is to shift from a fish community indicative of eutropy to a self-sustaining community indicative of mesotrophy. Quantitative fishery targets include (Hamilton Harbour Remedial Action Plan Writing Team 1992):

- 200-250 kg/ha total biomass of fish in littoral habitats
- 40-60 kg/ha piscivore biomass in littoral habitats
- 70-100 kg/ha specialist biomass in littoral habitats
- 30-90 kg/ha generalist biomass in littoral habitats
- native piscivores representing 20-25% of total biomass
- 80-90% native species
- a species richness of 6-7 species per survey transect.

▼ *Appendix B*

Use impairment	Listing and delisting guidelines	Example of quantitative objective/target for use restoration
Fish tumours or other deformities	*Listing Guideline:* When the incidence rates of fish tumours or other deformities exceed rates at unimpacted control sites or when survey data confirm the presence of neoplastic or preneoplastic liver tumours in bullheads or suckers. *Delisting Guideline:* When the incidence rates of fish tumours or other deformities do not exceed rates at unimpacted control sites and when survey data confirm the absence of neoplastic or preneoplastic liver tumours in bullheads or suckers.	In the Black River (Ohio), PAH contamination is known to cause fish tumours. Based on standardized fish survey techniques, two targets apply: no neoplastic liver tumours in a minimum sample of 25 brown bullhead ($\geq$ two years old); and the incidence rate of skin and lip tumours must be less than the incidence rate at a control site. One hundred and fifty control site and 130 contaminated site fish would be needed to verify a 5% difference (2% vs. 7%; 95% confidence) (Bauman 1992).
Bird or animal deformities or reproductive problems	*Listing Guideline:* When wildlife survey data confirm the presence of deformities (e.g., cross-bill syndrome) or other reproductive problems (e.g., egg-shell thinning) in sentinel wildlife species. *Delisting Guideline:* When the incidence rates of deformities (e.g., cross-bill syndrome) or reproductive problems (e.g., egg-shell thinning) in sentinel wildlife species do not exceed background levels in inland control populations.	In the lower Green Bay and Fox River Area of Concern (Wisconsin), historical discharges from the largest concentration of pulp and paper mills in the world are believed to have been the primary source of the 30,000 kg of PCBs that now reside in the sediments of the river downstream of Lake Winnebago and up to 15,000 kg of PCBs in Green Bay. Studies have demonstrated avian exposure to contaminants through aquatic food chains. A 1983 study of two colonies of Forster's tern showed the reproductive success of a lower Green Bay colony to be significantly impaired when compared to a relatively clean reference colony on Lake Poygan, upstream from industrial activities on the Fox River. Based on the 1983 study and an additional study in

1988, reproductive success was defined as a hatching rate of 90% based on the mean hatchability of the 1983 reference colony at Lake Poygan (Kubiak et al. 1989) and the mean hatchability of 155 populations of 113 avian species (Koenig 1982); a mean fledging rate of between 1.0 chick/pair measured at the 1983 reference colony; an average incubation time of 23 days; and a normal growth rate of chicks (body weight and length of wing, tarsus, bill, and head) based on 1988 data for chicks known to have successfully fledged form the Green Bay colony (Harris et al. 1993).

| Degradation of benthos | *Listing Guideline:* When benthic macroinvertebrate community structure significantly diverges from unimpacted control sites of comparable physical and chemical characteristics. In addition, this use will be considered impaired when toxicity (as defined by relevant, field-validated bioassays with appropriate quality assurance/quality controls) of sediment-associated contaminants at a site is significantly higher than controls.

Delisting Guideline: When benthic macroinvertebrate community structure does not significantly diverge from unimpacted control sites of comparable physical and chemical characteristics. Further, in the absence of community structure data, this use will be considered restored when toxicity of sediment-associated contaminants is not significantly higher than controls. | In Canada, site-specific guidelines for benthos are being established from a reference site data base (i.e., biological attributes and environmental variables) using multivariate techniques, such as cluster and ordination analysis (Reynoldson and Zarull 1993). Reference site benthic communities are grouped using cluster analysis. The site environmental variables, which are not affected or minimally affected by anthropogenic activity, are then used as predictors to group the sites into the appropriate biological clusters. The benthic community structure and the same nine environmental variables (depth, NO^3, silt, aluminum, calcium, loss on ignition, alkalinity, sodium, pH) are measured at the test sites. Using the environmental predictors and the discriminant model (derived from the reference site data base), each site is assigned to a biological cluster. The benthic invertebrate data are then similarly analyzed. If the observed (biological community) cluster lies outside the predicted cluster, then the site is judged to be impaired. |

▼ *Appendix B*

Use impairment	Listing and delisting guidelines	Example of quantitative objective/target for use restoration
		In the Great Lakes, 335 sites have been sampled and the multivariate "model" developed from this data base correctly predicts biological site clustering with 90% accuracy (Reynoldson et al. 1994). In addition, acute and chronic measures of "toxicity" (including growth and reproduction) are performed at these same sites, to provide measures of background performance and for the appropriate, indigenous organisms that are to be used in assessing sediment toxicity (see below).
Restrictions on dredging activities	*Listing Guideline:* When contaminants in sediments exceed standards, criteria, or guidelines such that there are restrictions on dredging or disposal activities. *Delisting Guideline:* When contaminants in sediments do not exceed standards, criteria, or guidelines such that there are restrictions on dredging or disposal activities.	Great Lakes dredging guidelines were developed to provide protection against the short- and long-term impacts associated with the disposal of dredged sediments. These guidelines employ bulk chemistry measurements for a few parameters that are assessed using either water quality equivalent standards or background concentration classifications (Zarull and Reynoldson 1992; UC 1982). More recently, the Ontario Ministry of Environment and Energy has released biologically-based sediment contaminant concentration guidelines for use in assessing bottom sediments in Areas of Concern and for use in assessing dredged material disposal. These chemical concentration guidelines are also supported through the use of site-specific bioassays (OMOE 1992).

In many areas outside the Great Lakes, the Sediment Quality Triad Approach (i.e., chemistry, benthos community structure, and bioassays) is being used to assess sediment problems and recommend remedial actions (Chapman 1990). A similar method has been recommended for use in the Great Lakes (UC 1987, 1988; Zarull and Reynoldson 1992).

Endpoints for benthos community structure are being established as described above, using reference sites throughout the nearshore Great Lakes. Sediment bioassays, an essential adjunct, provide confirmation that sediment is the source of the impact, rather than the water column or other factors, which are integrated by the benthos. As with community structure, a reference site (bioassay) data base has been established (Day et al. unpublished report). Examples of quantitative endpoints for standard sediment bioassays performed at "clean" sites include:

Chironomus riparius 10-day bioassay: 84% survival in all sediments and growth of 0.34 mg dry weight per individual;

Hexagenia limbata 21-day bioassay: 95% survival in all sediments, growth of 2.3 mg dry weight per individual in sand ($\geq$ 25%), and growth of 8.0 mg dry weight in silt ($\geq$ 40%);

Hyallella azteca 28-day bioassay: 44% survival in $\leq$ 20% silt, 88% survival in >20% silt, and growth of 0.48 mg dry weight in all sediments; and

Appendix B

Use impairment	Listing and delisting guidelines	Example of quantitative objective/target for use restoration
		Tubifex tubifex 28-day bioassay: nine cocoons per adult in all sediments and 24 young per adult in all sediments.
		If the community criteria (CC) and the bioassay criteria (BC) are met, then open water disposal is acceptable. If neither CC nor BC are met, then confinement and/or treatment are necessary. If CC are not met, but all BC are, then open water disposal is possible since community problem is not likely sediment related. If CC are not met but some BC are, then open water disposal is dependent upon the degree of acceptable risk. If CC are met, but some BC are not, then open water disposal is possible since the problem is not likely contaminant related. If CC are met, but all of the BC are not, then a careful reassessment of methods/procedures is required (this could also be a result of a highly adapted indigenous community).
Eutrophication or undesirable algae	*Listing Guideline:* When there are persistent water quality problems (e.g., dissolved oxygen depletion of bottom waters, nuisance algal blooms or accumulation, decreased water clarity, etc.) attributed to cultural eutrophication.	In Saginaw Bay, Lake Huron, modelling phosphorous loading-phosphorous concentration-threshold odour value relationships has led to establishment of a 15 mg/L total phosphorous (TP) concentration for the inner bay (Bierman et al. 1983). The TP loading target is 440 tonnes/yr, which will result in threshold odour values <3 and a TP concentration of 15 mg/L.

In Green Bay, Lake Michigan, regression analysis has been used to model the relationships among TP loading, TP concentration, total suspended solids, chlorophyll *a*, and water clarity. Based on a 0.7 m Secchi depth (summer average) necessary to restore a submerged aquatic vegetation (McAllister 1991), trophic state objectives were established as follows: 90 ug/L summer average TP, 25 ug/L summer average chlorophyll *a*, and 10 mg/L total suspended solids. These values correspond to an annual TP load of about 350 tonnes/yr, or a 50% reduction in current loading (WDNR 1993).

Delisting Guideline: When there are no persistent water quality problems (e.g., dissolved oxygen depletion of bottom waters, nuisance algal blooms or accumulation, decreased water clarity, etc.) attributed to cultural eutrophication.

Restrictions on drinking water consumption or taste or odour problems

Listing Guideline: When treated drinking water supplies are impacted to the extent that (1) densities of disease causing organisms or concentrations of hazardous/toxic chemicals or radioactive substances exceed human health standards, objectives, or guidelines, (2) taste and odour problems are present, or (3) treatment needed to make raw water suitable for drinking is beyond the standard treatment used in comparable portions of the Great Lakes that are not degraded (i.e., settling, coagulation, disinfection).

Delisting Guideline: For treated drinking water supplies: (1) when densities of disease causing organisms or concentrations of hazardous/toxic chemicals or radioactive substances do not exceed human health standards, objectives, or guidelines,

In the Maumee River Area of Concern in southwestern Lake Eric, nitrate levels have increased above 10 mg/L during spring and fall in some municipal water supplies. When this occurs, drinking water consumption warnings are issued because elevated levels of nitrate have been found to be harmful to certain groups of people (e.g., excessive nitrate causes methemoglobinemia in infants). Drinking water consumption warnings are removed by the municipalities when nitrate levels fall below 10 mg/L for two consecutive days based on standardized sampling and analytical techniques.

In Saginaw Bay, Lake Huron, taste and odour problems associated with blue-green algae have been identified in the municipal water supplies. Threshold odour is quantitatively measured and ranked on a scale from one to ten based on the dilution necessary to ensure that taste and odour are

Use impairment	Listing and delisting guidelines	Example of quantitative objective/target for use restoration
	(2) when taste and odour problems are absent, and (3) when treatment needed to make raw water suitable for drinking does not exceed standard treatment as defined above.	barely detectable, with a value of three being the U.S. Public Health Service Threshold Standard (Bierman et a. 1983). Threshold odour is measured daily and biweekly averages are calculated to determine compliance with the U.S. Public Health Service Standard of three.
Beach closings	*Listing Guideline:* When waters that are commonly used for total body-contact or partial body-contact recreation exceed standards, objectives, or guidelines for such use.	Along the Metropolitan Toronto Waterfront (Lake Ontario), numerous beaches are posted unsafe for swimming as a result of high bacterial counts from stormwater runoff and combined sewer overflows. The Ontario Ministry Health Standard is 100 colonies *Escherichia coli*/100ml. Beaches are considered safe for swimming when the daily geometric mean of a minimum of five samples collected from different sites within the beach area is less than 100 conolies/100 ml based on standardized sampling protocols (Ontario Ministry of Health 1992).
	Delisting Guideline: When waters that are commonly used for total body-contact or partial body-contact recreation do not exceed standards, objectives, or guidelines for such use.	In Wisconsin, both narrative and numerical standards are set for public swimming beaches. Waters must be free of chemical substances capable of creating toxic reactions or irritations to skin/membranes, must achieve numerical bacterial standards, and must achieve a 4 m Secchi Disc water clarity standard for safety reasons (Wisconsin Adm. Rule HSS 171).

| Degradation of aesthetics | *Listing Guideline:* When any substance in water produces a persistent objectionable deposit, unnatural colour or turbidity, or unnatural odour (e.g., oil slick, surface scum).

Delisting Guideline: When the waters are devoid of any substance that produces a persistent objectionable deposit, unnatural colour or turbidity, or unnatural odour (e.g., oil slick, surface scum). | In New York, narrative standards for suspended sediment and colour are set at "none" that would adversely affect the waters for their best use (New York State 1991). For turbidity, the standard is no increase that would cause a visible contrast from natural conditions and, for oil and floating substances, it is no residue that would be visible. If conditions are attributable to unnatural causes and sources, New York ambient water quality standards are used to establish reduction targets in order to make a determination. Examples of quantitative targets that have been established for dischargers causing such conditions include: 3.0 mg/L for suspended solids; and 15 mg/L for oil and floating substances. |
| Added costs to agriculture or industry | *Listing Guideline:* When there are additional costs required to treat the water prior to use for agricultural purposes (i.e., including, but not limited to, livestock watering, irrigation, and crop-spraying) or industrial purposes (i.e., intended for commercial or industrial applications and noncontact food processing).

Delisting Guideline: When there are no additional costs required to treat the water prior to use for agricultural or industrial purposes (as defined above). | In the St. Clair River Area of Concern, "added costs to agriculture or industry" have been identified as an impaired beneficial use. Food processing industries in Ontario and a salt processes facility in Michigan had to temporarily shut down their intakes due to upstream spills in 1990 and 1989, respectively (Ontario Ministry of the Environment and Michigan Department of Natural Resources 1991). In both instances, added costs to these industries were approximately $2,000/hour during the spill events. This use is considered restored when there are no added costs to treat the water prior to use in industrial or agricultural processes. |

▼ Appendix B

Use impairment	Listing and delisting guidelines	Example of quantitative objective/target for use restoration
Degradation of phytoplankton and zooplankton populations	*Listing Guideline:* When phytoplankton or zooplankton community structure significantly diverges from unimpacted control sites of comparable physical and chemical characteristics. In addition, this use will be considered impaired when relevant, field-validated phytoplankton or zooplankton biossays (e.g., *Ceriodaphnia*, algal fractionation bioassays) with appropriate quality assurance/quality controls confirm toxicity in ambient waters. *Delisting Guideline:* When phytoplankton or zooplankton community structure does not significantly diverge from unimpacted control sites of comparable physical and chemical characteristics. Further, in the absence of community structure data, this use is considered restored when plankton bioassays confirm no toxicity in ambient waters.	Limited attempts have been made to quantify objectives based on zooplankton and phytoplankton enumeration and quantification. Bioassay endpoints are more frequently used. Degraded zooplankton populations were identified as an impaired use in the Cuyahoga River due to chronic toxicity of ambient waters below the Akron Wastewater Treatment Plant. Toxicity was measured be the seven-day, three brood *Ceriodaphnia* test. *Ceriodaphnia* are easily cultured, found in the Great Lakes, sensitive to toxic substances, and have a short maturation time. Based on standard *Ceriodaphnia* bioassay protocols (UC 1987), zooplankton populations were considered not impaired when there was no significant difference in survival and number of young per female relative to controls ($P < 0.05$).
Loss of fish and wildlife habitat	*Listing Guideline:* When fish and wildlife management goals have not been met as a result of loss of fish and wildlife habitat due to a perturbation in the physical, chemical, or biological integrity of the Boundary Waters, including wetlands.	Approximately 80% of the wetlands in Hamilton Harbour, Lake Ontario, have been lost to development. The water use goal for fishery is "that water quality and fish habitat should be improved to permit an edible, naturally reproducing fishery for warm water species, and water and

Delisting Guideline: When the amount of physical, chemical, and biological habitat required to meet fish and wildlife management goals has been achieved and protected.

habitat conditions in Hamilton Harbour should not limit natural reproduction and the edibility of cold water species." This water use goal has been translated into the following targets for fish habitat (Hamilton Harbour Remedial Action Plan Writing Team 1992): increase the quantity of emergent and submergent aquatic plants in the Hamilton Harbour, Cootes Paradise, Grindstone Creek Delta, and Grindstone Creek marshes to approximately 500 ha in accordance with the Fish and Wildlife Habitat Restoration Project; rehabilitate 344 ha of littoral fish habitat; rehabilitate 39 ha of pike spawning marsh and nursery habitat; provide additional 10 km of littoral shore by creating 5 km of narrow islands; and achieve water clarity as measured by Secchi Disc during the summer season of 3.0 m in the Harbour and 1.0 m in the Cootes Paradise and Grindstone Creek.

Source: Modified from Hartig, Hartig, and Laws (1994, 344-48). For references cited above, see Hartig, Hartig, and Laws (1994, 352).

References

Ali, A. 2000. "Multistakeholder Collaborative Planning for Common Property Resources." Unpublished paper, McMaster University, Hamilton.

Ashworth, W. 1986. *The Late Great Lakes*. New York: Knopf.

Bagehot, W. 1964. *The English Constitution*. London: Watts and Co.

Benson, B. 1990. *The Enterprise of Law*. San Francisco: Pacific Research Institute for Public Policy.

Berman, H.J. 1983. *Law and Revolution: The Formation of the Western Legal Tradition*. Cambridge, MA: Harvard University Press.

Berton, P. 1996. *The Great Lakes*. Toronto: Stoddart.

Bixby, A. 1986. *The Law and the Lakes*. Toronto: Centre for the Great Lakes.

Bixby, A.A. 1985. "The Great Lakes Water Quality Agreement and Areas of Concern." MA thesis, University of Toronto.

Boyle, D.E. 1990. *Weighing the Options*. Halifax: Oceans Institute of Canada.

Breton, A., and R. Wintrobe. 1979. "Bureaucracy and State Intervention." *Canadian Public Administration* 22: 208-26.

Bromley, D.W., D. Feeny, M.A. McKean, P. Peters, J.C. Gilles, R.J. Oakerson, C.F. Runge, and J.T. Thomson, eds. 1992. *Making the Commons Work: Theory, Practice, and Policy*. San Francisco: Institute for Contemporary Studies Press.

Buchanan, J.M., and G. Tullock. 1962. *The Calculus of Consent*. Ann Arbor: University of Michigan Press.

Cairns, R.D. 1992. "Natural Resources and Canadian Federalism: Decentralization, Recurrent Conflict, and Resolution." *Publius* 22: 55-71.

Caldwell, L.K. 1963. "Environment: A New focus for Public Policy." *Public Administration Review* 23: 132-39.

–, ed. 1988. *Perspectives on Ecosystem Management for the Great Lakes*. Albany, NY: State University of New York Press.

Campbell, R., A.D. Scott, P. Pearse, and M. Uzelac. 1974. "Water Management in Ontario." *Osgoode Hall Law Journal* 12: 475-526.

Canada. Department of Fisheries and Oceans (DFO). 1999. "1995 Selected Results for the Great Lakes Fishery." *Survey of Recreational Fishing in Canada*. Ottawa: DFO.

Canada-Ontario RAP Steering Committee. 1995. *Updates in Progress of Canada-Ontario Remedial Action Plans*. Toronto: Environment Canada.

Chandler, J.G., and M.J. Vechsler. 1992. "The Great Lakes: St. Lawrence River Basin from an IJC Perspective." *Canadian-United States Law Journal* 18: 261-85.

Coase, Richard H. 1960. "The Problem of Social Cost." *Journal of Law and Economics* 3: 1-44.

Colborn, T., A. Davidson, S.M. Green, R.A. Hodge, C.I. Jackson, and R.A. Liroff. 1990. *Great Lakes, Great Legacy?* Washington, DC: Conservation Foundation; Ottawa: Institute for Research on Public Policy.

Commons, John R. 1924. *The Legal Foundations of Capitalism.* Madison: University of Wisconsin Press.

Davies, J.C. 1970. *The Politics of Pollution.* New York: Pegasus.

de Groot, R. 1986. *A Functional Ecosystem Evaluation Method as a Tool in Environmental Planning and Decision Making.* Wageninen, The Netherlands: Wageninen Agricultural University.

Environment Canada 2000. *Detroit River Remedial Action Plan, 1998.* <www.cciw.ca/glimr/raps/connecting/detroit>.

Doern, G.B., and T. Conway. 1994. *The Greening of Canada.* Toronto: University of Toronto Press.

Doherty, R. 1990. *Disputed Waters: Native Americans and the Great Lakes Fishery.* Lexington: University of Kentucky Press.

Dubin, J.A. 1974. *Jackson* v. *Drury Construction Co. Ltd.* Ottawa: Revised Statutes of Canada, 186.

Dunleavy, P. 1991. *Politicians, Bureaucrats and Democracy.* London: Harvester: Wheatsheaf.

Dwivedi, O.P. 1974. *Protecting the Environment.* Toronto: Copp Clark.

–. 1980. *Resources and Environment.* Toronto: McClelland and Stewart.

Ecowise. 1996. *McMaster Eco-Research Program for Hamilton Harbour: Final Report, 1993-96.* Hamilton, ON: McMaster University.

Eiger, N., and P. McAvoy. 1992. *Empowering the Public.* Chicago: Center for the Great Lakes and the EPA.

Environment Canada. 1997. *Lakewide Management Plan for Lake Ontario* Burlington, ON: Environment Canada.

Environment Canada and United States Environmental Protection Agency. 1995a. *Environmental Atlas and Resource Book: The Great Lakes.* Ottawa: Environment Canada; Washington, DC: Environmental Protection Agency.

–. 1995b. *State of the Great Lakes 1995.* Burlington, ON/Chicago, IL: Environment Canada/Environmental Protection Agency.

–. 1997. *Canada-United States Strategy for the National Elimination of Persistent Toxic Substances in the Great Lakes: The Great Lakes Binational Toxics Strategy.* Toronto/Chicago: Environment Canada/Environmental Protection Agency.

Environment Canada, United States Environmental Protection Agency, Ontario Ministry of Environment and Energy, and New York Department of Environmental Conservation. 1995. *Joint Evaluation of Upstream/Downstream Niagara River Monitoring Data.* Toronto: Environment Canada.

Environmental Protection Agency. 1993. *LaMP Fact Sheet.* Chicago: EPA Great Lakes National Program Office.

Feldman, E.J., and M.A. Goldberg, eds. 1987. *Land Rites and Wrongs.* Lincoln, NE: Lincoln Institute of Land Policy.

Filyk, G., and R. Coté. 1992. "Pressure from the Inside: Advisory Groups and the Environmental Policy Community." In R. Boardman, ed., *Canadian Environmental Policy*, 60-82. Toronto: Oxford University Press.

Fiorino, D.J. 1995. *Making Environmental Policy*. Berkeley: University of California Press.

Francis, G. 1994. "Ecosystems." Paper presented at Social Sciences Federation of Canada Conference, Social Sciences and the Environment, 17-19 February, Ottawa.

Freeman, M.A. 1990. "Water Pollution Policy." In D.R. Portney, ed., *Public Policies for Environmental Protection*, 97-149. Washington, DC: Resources for the Future.

Gebhardt, A., and G. Lindsey. 1996. "Financing the Ecosystem Approach." *Public Works Management and Policy* 1: 158-73.

Gibson, R.B., and B. Savan. 1980. *Environmental Assessment in Ontario*. Toronto: Canadian Environmental Law Research Foundation.

Great Lakes Fishery Commission. 1979. *Commercial Fish Production in the Great Lakes, 1967-1977*. Technical Reports. Ann Arbor, MI: GLFC.

–. 1993. *Strategic Vision of the Great Lakes Fishery Commission for the Decade of the 1990's*. In Great Lakes Fishery Commission, Lake Erie Committee, 1993 Annual Meeting, Appendix XIV. Ann Arbor, MI: Great Lakes Fishery Commission.

Great Lakes Water Quality Board (GLWQB). 1985. *1985 Report on Great Lakes Water Quality*. Windsor: International Joint Commission.

–. 1987. *1987 Report on Great Lakes Water Quality*. Windsor: International Joint Commission.

–. 1991a. *Review and Evaluation of the Great Lakes RAP Program*. Windsor: International Joint Commission.

–. 1991b. *International Joint Commission Review of Remedial Action Plans for the Great Lakes Areas of Concern*. Windsor: International Joint Commission.

Gunderson, L.H., C.S. Holling, and S.S. Light, eds. 1995. *Barriers and Bridges to the Renewal of Ecosystems and Institutions*. New York: Columbia University Press.

Gunther-Zimmerman, A. 1994. "Ecosystem Approach to Planning in the Great Lakes." PhD dissertation. University of Toronto.

Harrison, K., and G. Hoberg. 1996. *Risk, Science, and Politics: Regulating Toxic Substances in Canada and the United States*. Montreal: McGill-Queen's University Press.

Hartig, J.H., and N.L. Law. 1994. *Progress in Great Lakes Remedial Action Plans*. Detroit: Wayne State University.

Hartig, J.H., and R.L. Thomas. 1988. "Development of Plans to Restore Degraded Areas in the Great Lakes." *Environmental Management* 12: 327-47.

Hartig, J.H., and M.A. Zarull, eds. 1992. *Under RAPs: Toward Grassroots Ecological Democracy in the Great Lakes Basin*. Ann Arbor: University of Michigan Press.

Hartig, J.H., P. Hartig, and B. Laws. 1994. "Quantifying 'How Clean Is Clean?' for Degraded Areas in the Great Lakes." Paper presented at Water Environment Federation, 67th Annual Conference and Exposition, Chicago, IL, 15-19 October, 341-52.

Hartig, J.H., D.P. Dodge, D. Jester, J. Atkinson, R. Thoma, and K. Cullis. 1996. "Towards Integrating Remedial Action Planning and Fishery Management Planning in Great Lakes Areas of Concern." *Fisheries* 21: 6-13.

Hartig, J.H., N.L. Law, D. Epstein, K. Fuller, J. Letterhos, and G. Krantzberg. 1995. "Capacity Building for Restoring Degraded Areas in the Great Lakes." *International Journal of Sustainable Development and World Ecology* 2: 1-10.

Hoberg, G. 1992. "Comparing Canadian Performance in Environmental Policy." In R. Boardman, ed., *Canadian Environmental Policy*, 246-62. Toronto: Oxford University Press.

–. 1993. "Environmental Policy: Attractive Styles." In M.M. Atkinson, ed., *Governing Canada*, 307-42. Toronto: Harcourt, Brace, Jovanovich.

Hohfeld, Wesley. 1919. *Fundamental Legal Conceptions.* New Haven: Yale University Press.

Holling, C.S. 1986. "Resilience of Ecosystems: Local Surprise and Global Change." In W.C. Clark and R.E. Munn, eds., *Sustainable Development of the Biosphere*, 292–317. Cambridge: Cambridge University Press.

Holmes, J., and T.H. Whillans. 1984. *Historical Review of Hamilton Harbour Fisheries.* Canadian Technical Report of Fisheries and Aquatic Sciences No. 1257. Burlington, ON: Fisheries and Oceans Canada.

Howlett, M. 1994. "The Judicialization of Canadian Environmental Policy, 1980-90." *Canadian Journal of Political Science* 27: 88-128.

Ingram, H., and D.E. Mann. 1989. "Interest Groups and Environmental Policy." In J.P. Lester, ed., *Environmental Politics and Policy*, 135–57. London: Duke University Press.

Inscho, F.R., and M.H. Durfee. 1995. "The Troubled Renewal of the Canada-Ontario Agreement Respecting Great Lakes Water Quality." *Publius* 25: 51-69.

International Joint Commission. 1989. *Great Lakes Water Quality Agreement of 1978, Revised.* Windsor, ON: IJC.

–. 1991a. *Review and Evaluation of the Great Lakes Remedial Action Plan.* Windsor, ON: IJC.

–. 1991b. "Commission Approves List/Delist of Criteria for Great Lakes Areas of Concern." *Focus* 15 (March/April): 4-5.

–. 1994. *Revised Great Lakes Water Quality Agreement of 1978.* Windsor, ON: IJC.

–. 1997a. *Detroit River Area of Concern: Status Assessment.* Windsor, ON: IJC.

–. 1997b. *Overcoming Obstacles to Sediment Remediation in the Great Lakes Basin.* Windsor, ON: Great Lakes Water Quality Board. <www.ijc.org>.

–. 1998. *Beacons of Light.* Windsor, ON: Great Lakes Water Quality Board <www.ijc.org>.

–. 2000. *Protection of the Waters of the Great Lakes.* Windsor, ON: IJC.

Johns, C. 2000. "Non-Point Source Water Pollution Management in Canada and the United States." PhD dissertation. McMaster University, Hamilton, ON.

Kaufman, K. 1989. *Of the Great Lakes in the Seventeenth Century.* Providence, RI: John Carter Brown Library.

Kettl, D. 1993. *Sharing Power.* Washington: Brookings Institution.

Knox, L. 1999. *Remedial Action Plan for Hamilton Harbour: 1998 Status Report.* Burlington, ON: Hamilton Harbour RAP Office.

Krantzberg, G., H. Ali, and J. Barnes. 1998. *Incremental Programs in Restoring Beneficial Uses at the Canadian Areas of Concern.* Toronto: Ontario Ministry of Environment.

LaForest, Gerard V. 1969. *Natural Resources and Public Property under the Canadian Constitution.* Toronto: University of Toronto Press.

Lambden, D.W., and Izaak De Rijcke. 1985. "Boundaries and Surveys." Title 19 in *Canadian Encyclopedic Digest*. Toronto: Carswell.

Lee, Kai N. 1993. *Compass and Gyroscope*. Washington, DC: Island Press.

Lester, J.P. 1986. "New Federalism and Environmental Policy." *Publius* 16: 149-65.

Levine, A.G. 1982. *Love Canal*. Lexington, MA: Lexington Books.

Libecap, G.B. 1986. "Property Rights in Economic History." *Explanations in Economic History* 23: 227-52.

Lucas, A. 1990. "The New Environmental Law." In R.J. Watts and D. Brown, eds., *Canada: State of the Federation*, 167-92. Kingston, ON: Queen's University.

MacKenzie, S.H. 1996. *Integrated Resource Planning and Management*. Washington, DC: Island Press.

Marcus, A.A. 1980. *Promise and Performance*. New Haven: Greenwood Press.

McCalla, R.J. 1994. *Water Transportation in Canada*. Halifax: Formac.

McCay, B.J., and J.M. Acheson. 1987. *The Question of the Commons*. Tucson: University of Arizona Press.

McKean, M.A. 1992. "Success in the Commons." *Journal of Theoretical Politics* 4: 247-82.

Mitchell, B., and D. Shrubsole. 1992. *Ontario Conservation Authorities: Myth and Reality*. Geography Publication Series 35. Waterloo, ON: University of Waterloo.

Muldoon, P.R. 1983. "The International Joint Commission and Point Roberts." MA thesis. McMaster University, Hamilton, ON.

Munton, D. 1980. "Great Lakes Water Quality." In O.P. Dwivedi, ed., *Resources and the Environment*, 153-78. Toronto: McClelland and Stewart.

Wayne State University, Department of Civil Engineering. 1994. *Institutional Frameworks to Direct Ongoing Development and Implementation of RAPS*. Detroit: Wayne State University.

Nelles, H.V. 1974. *The Politics of Development*. Toronto: Archon.

New York State, Department of Environmental Conservation. 1994. US Niagara River RAP, Stage 1, 4-99.

Niskanen, W.A. 1971. *Bureaucracy and Representative Government*. New York: Aldine-Atherton.

Organization for Economic Co-operation and Development. 1996. *Environmental Performance Reviews*. Paris: OECD.

Olson, M. 1965. *The Logic of Collective Action*. New Haven: Yale University Press.

Ostrom, E. 1990. *Governing the Commons*. New York: Cambridge University Press.

Ostrom, E., R. Gardner, and J. Walker. 1994. *Rules, Games and Common Pool Resources*. Ann Arbor: University of Michigan Press.

Ostrom, V. 1980. "Artisanship and Artifact." *Public Administration Review* 40: 309-17.

–. 1989. *The Intellectual Crisis in American Public Administration*. 2nd ed. Tuscaloosa: University of Alabama Press.

–. 1991. *The Meaning of American Federalism*. San Francisco: Institute for Contemporary Studies.

–. 1997. *The Meaning of Democracy and the Vulnerability of Democracies*. Ann Arbor: University of Michigan Press.

Pearse, P., F. Bertrand, and J.W. MacLaren. 1985. *Currents of Change: Final Report of the Inquiry on Federal Water Policy*. Ottawa: Environment Canada.

Pinkerton, E., ed. 1989. *Cooperative Management of Local Fisheries*. Vancouver: University of British Columbia Press.

Prothero, F. 1973. *The Good Years: A History of the Commercial Fishing Industry on Lake Erie.* Belleville, ON: Mika Publishing.

Rabe, B.G., and J.B. Zimmerman. 1995. "Beyond Environmental Regulatory Fragmentation." *Governance* 8: 58-77.

Regier, Henry A. 1986. "Programs with Remediation, Rehabilitation, and the Ecosystem Approach." *Alternatives* 13, 3: 51.

Ringquist, R. 1993. *Environmental Protection at the State Level.* New York: M.E. Sharpe.

Rogers, P. 1996. *America's Water.* A Twentieth Century Fund Book. Cambridge, MA: MIT Press.

Rosenbaum, W. 1991. *Environmental Politics and Policy.* 2nd ed. Washington, DC: Congressional Quarterly Press.

Royal Society of Canada and the National Research Council of the United States. 1985. *The Great Lakes Water Quality Agreement.* Washington, DC: National Academy Press.

St. Lawrence Seaway Authority. 1971, 1992. *Annual Reports.* Ottawa, ON.

Schlager, Edella, and E. Ostrom. 1992. "Property Rights Regimes and Natural Resources." *Land Economics* 68: 249-62.

Skogstad, G., and P. Kopas. 1992. "Environmental Policy in a Federal System." In R. Boardman, ed., *Canadian Environmental Policy,* 43-59. Toronto: Oxford University Press.

Smith, Vernon L. 1993. "Humankind in Prehistory." In Terry Anderson and Randy T. Simmons, *The Political Economy of Customs and Culture,* Chapter 9, 157-84. Lanham, MD: Rowman and Littlefield.

Sproule-Jones, M.H. 1974. "Towards a Dynamic Analysis of Collective Action." *Western Political Quarterly* 26: 414-26.

–. 1981. *The Real World of Pollution Control.* Vancouver: Westwater Research Centre, University of British Columbia.

–. 1984. "The Enduring Colony." *Publius* 14: 93-108.

–. 1993. *Governments at Work.* Toronto: University of Toronto Press.

–. 2000. "Horizontal Management." *Canadian Public Administration* 43: 91-109.

Sproule-Jones, M.H., and P.L. Richards. 1984. "Toward a Theory of the Regulated Environment." *Canadian Public Policy* 10: 305-15.

State of the Lakes Ecosystem Conference (SOLEC), T.A. Edsall, and M.N. Charlton. 1996a. "Nearshore Waters of the Great Lakes." Working Paper, State of the Lakes Ecosystem Conference. Ottawa and Washington: Environment Canada and US EPA.

State of the Lakes Ecosystem Conference (SOLEC), A. Thorp, R. Rivers, V. Pebbles. 1996b. "Impacts of Changing Land Use." Working Paper, State of the Lakes Ecosystem Conference. Ottawa and Washington: Environment Canada and the US EPA.

State of the Lakes Ecosystem Conference (SOLEC), R. Reid, and K. Holland. 1996c. "The Land by the Lakes." Working Paper, State of the Lakes Ecosystem Conference. Ottawa and Washington: Environment Canada and the US EPA.

Strutt, A. 1993. *Citizen Participation in Water Quality Management.* BA (Honours) thesis. McMaster University, Hamilton, ON.

Sussman, G. 1979. *Quebec and the St. Lawrence Seaway.* Montreal: CD Howe Institute.

Talhelm, D.R. 1988. *Economics of Great Lakes Fisheries.* Ann Arbor: Great Lakes Fishery Commission.

Thompson, J.D. 1969. *Organizations in Action*. New York: McGraw Hill.

Toro, C.E., and L.P. Dowd. 1961. *The St. Lawrence Seaway*. Ann Arbor: University of Michigan Press.

United States Environmental Protection Agency and Government of Canada. 1995. *The Great Lakes*. 3rd ed. Toronto: Government of Canada.

Vining, A. 1981. "Provincial Hydro Utilities." In A. Tupper and G.B. Doern, eds., *Public Corporations and Public Policy in Canada*, 152-75. Montreal: Institute for Research on Public Policy.

Wallace, F.W. 1945. *Canadian Fisheries Manual*. Gardenwall: National Business Publications.

Weiner, L.E. 1997. "Private Property: Conceptual and Normative Analysis." PhD dissertation, Harvard University, Cambridge, MA.

Wilson, J. 1992. "Green Lobbies: Pressure Groups and Environmental Policy." In R. Boardman, ed., *Canadian Environmental Policy*, 109-25. Toronto: Oxford University Press.

Index

Set in Stone by Artegraphica Design
Printed and bound in Canada by Friesens
Indexer: Pat Buchanan
Cartographer: Eric Leinberger